Resolving Development Disputes Through Negotiations

ENVIRONMENT, DEVELOPMENT, AND PUBLIC POLICY
A series of volumes under the general editorship of
Lawrence Susskind, *Massachusetts Institute of Technology, Cambridge, Massachusetts*

ENVIRONMENTAL POLICY AND PLANNING
Series Editor:
Lawrence Susskind, *Massachusetts Institute of Technology, Cambridge, Massachusetts*

CAN REGULATION WORK?
Paul A. Sabatier and Daniel A. Mazmanian

PATERNALISM, CONFLICT, AND COPRODUCTION
Learning from Citizen Action and Citizen Participation in
Western Europe
Lawrence Susskind and Michael Elliott

BEYOND THE NEIGHBOHOOD UNIT
Residential Environments and Public Policy
Tridib Banerjee and William C. Baer

RESOLVING DEVELOPMENT DISPUTES THROUGH
NEGOTIATIONS
Timothy J. Sullivan

ENVIRONMENTAL DISPUTE RESOLUTION
Lawrence S. Bacow and Michael Wheeler

Other subseries:

CITIES AND DEVELOPMENT
Series Editor:
Lloyd Rodwin, *Massachusetts Institute of Technology, Cambridge, Massachusetts*

PUBLIC POLICY AND SOCIAL SERVICES
Series Editor:
Gary Marx, *Massachusetts Institute of Technology, Cambridge, Massachusetts*

Resolving Development Disputes Through Negotiations

Timothy J. Sullivan

Graduate School of Public Policy
University of California, Berkeley
Berkeley, California

Plenum Press • New York and London

Library of Congress Cataloging in Publication Data

Sullivan, Timothy J., 1950–

 Resolving development disputes through negotiations.

 (Environment, development, and public policy. Environmental policy and planning)
 Bibliography: p.
 Includes index.
 1. Real estate development—Social aspects. 2. Negotiation in business. I. Title. II. Series.
HD1390.S94   1984   658   84-11696

ISBN-13: 978-1-4612-9705-5   e-ISBN-13: 978-1-4613-2757-8
DOI:10.1007/ 978-1-4613-2757-8

To My Parents

Preface

In the last decade, disputes between developers and local communities over proposed construction projects have led to increasing litigation. Environmental legislation, in particular, has greatly enhanced the rights and powers of organized groups that desire to participate in local development decisions. These powers have allowed citizen groups to block undesired and socially unacceptable projects, such as highways through urban areas and sprawling suburban developments. At the same time, these powers have produced a collective inability to construct many needed projects that produce adverse local impacts. Prisons, airports, hospitals, waste treatment plants, and energy facilities all face years of litigation before a final decision. At times, prolonged litigation has produced especially high costs to all participants.

Despite these new powers, citizen action has often been limited to participation in public hearings or adjudicatory proceedings. Typically, this occurs so late in the decision process that citizen input has very little affect in shaping a project's design. Those who dislike some element of a project often have little choice other than to oppose the entire project through litigation.

In sharp contrast to these often destructive litigation battles, a number of disputes between organized community groups and developers have reached a resolution through bargaining. These negotiations have produced settlements unavailable through traditional adversarial or adjudicatory proceedings. These agreements have included mitigation and compensation measures that have made projects acceptable to opposition groups while enabling everyone to avoid protracted adjudicatory proceedings. The success of these negotiation efforts leads one to wonder why bargaining takes place so rarely, and whether bargaining offers a process for participatory decision-making that can avoid the costly outcomes of adversarial adjudication.

This book argues that negotiation occurs so rarely because the existing procedures for reviewing proposed projects channel development conflicts into adjudication. The analysis compares the emerging structures of negotiations in development disputes to other systems of bargaining, principally collective bargaining and international negotiations. It concludes that the incentives acting on the participants, the organization and issue structures, and the bargaining environment cause negotiations over development disputes to resemble more closely those of international negotiations than those of industrial relations. Without the support of domestic law, negotiations over development disputes will lead to agreement only in those rare situations where leaders and group members clearly see the benefits of resolving a dispute through negotiation. As in international relations, successful negotiations over development disputes will prove a surprising exception rather than the standard rule.

Industrial relations, on the other hand, offer a system of bargaining where negotiations routinely produce agreements without prolonged strikes. Law structures a conflict and creates incentives that encourage the resolution of disputes through negotiation. Labor law determines who legitimately represents workers, exhorts negotiators to bargain in good faith, offers mediators to help resolve impasses, regulates the tactics of industrial conflict, and endorses agreements.

This book argues that only when laws provide similar support for negotiated development will the potential benefits of bargaining be realized. Laws can help support efforts to resolve disputes through negotiation by establishing procedures for selecting bargaining participants, by legitimating an agenda of issues, and by conferring a legal status on negotiated settlements. Without the support of law, negotiations can successfully avoid adjudication only in those rare situations where stalemates and high costs induce disputants to search for alternative ways to resolve a conflict.

Negotiations over development disputes pose special problems. Many of the participants bargain with each other for the first and only time. Community organizations often have little training in bargaining procedures. For these reasons, mediators have played and will continue to play a major role in bargaining efforts to resolve development disputes. The analysis of this book addresses the special issues raised by the participation of third parties in these disputes.

Good policies and laws seldom result from the ideas of one, but more often arise from a discussion among many who contribute fresh insights and facts. This book offers policymakers a conceptual framework for generating the questions that they need to ask in determining whether and how to use negotiations. The analysis of this text should prove especially helpful to legislators, government officials, and planning boards charged with the task of siting or reviewing unwanted but needed facilities. It should prove particularly helpful to those officials who must help forge a public consensus from legitimate but competing concerns.

Timothy J. Sullivan

Acknowledgments

This book grew from an ongoing discussion with the ideas and analyses of others. Although the text cites the writings of many who have studied negotiation processes, the shape of many of the ideas in this book resulted from more personal discussions. Lawrence Susskind stimulated this writing with his extensive knowledge of negotiations and with his editorial prods to deeper analyses. Martin Levin's criticisms helped transform the writing from a collection of insights into an argument. Howard Raiffa, my thesis adviser, supported the earliest formulations of these ideas both by constructively criticizing my graduate work and by arranging for financial support. Michael O'Hare and Eugene Bardach offered countless helpful insights and suggestions.

Many individuals have contributed to the ideas in this book. Lawrence Bacow, Henry Brady, Stephen Hill, Bart McGuire, Arnold Meltsner, Dail Phillips, John Quigley, Aaron Wildavsky, and Richard Zeckhauser deserve special mention. Andrea Altschuler added an undergraduate's fresh perspectives that helped ideas mature without ossifying. Classes of Berkeley undergraduates taught me both to entertain ideas playfully and to examine them critically. Karen Chin prepared many drafts of this manuscript while overcoming the technical problems posed by a new computer text-editor.

Institutions provide homes where individuals can share ideas. The Graduate School of Public Policy has provided an academic home that enabled me to write this book. A special postdoctoral program at the Massachusetts Institute of Technology's Department of Urban Studies and Planning introduced me to many of the individuals who have continued to discuss these ideas with me for several years. The Massachusetts Institute of Technology's Environmental Negotiation Project offered an opportunity to develop new ideas through fieldwork, and to share them with others. Harvard's Kennedy School of Government served as a first home for this work.

T.J.S.

Contents

Introduction

Life is full of negotiations. Even childhood work and play involve bargaining. Trading household chores with a brother or sister or trading toys with a playmate gives us early experience with the potential of bargaining for solving and creating problems. As one matures, one finds that being older, stronger, or smarter does not guarantee that one always gets one's way. Few people are born into a family where preferences are complementary or where rigid hierarchies require absolute obedience. If give-and-take do not rule the day, we nevertheless spend a great deal of time doing it. Reciprocal backscratching beats the contortions necessary for self-sufficiency. Even childhood experience suggests that nonnegotiable demands are intolerable. Sooner or later, these demands lead a sibling to passive resistance or to a more active fight. Although life has some situations where one only wins what another loses, more often reciprocal concessions leave all better off. The trick, of course, is to find the right combinations. Is washing dishes or profit sharing the better offer for a younger brother's help with the Sunday's newspaper deliveries? One has to know one's brother, or at least remain on speaking terms.

As one gets older, family disputes do not disappear. For many, they grow in intensity as family members pass through adolescence. Fights between a father and an adolescent son can provoke each into rigid positions. Each side quickly takes stands that brook no compromise or concession. Left alone, these fights can simmer for weeks and can leave no member of the family untouched, either with sympathy for those fighting, or with concern for the trouble that they impose on the rest of the family.

In a large enough family, some member with a distance from the immediate issues will act as a go-between. This mediator, even when

1

lacking the authority to resolve the dispute, helps to moderate the conflict and to restart conversation. Such a mediator can make effective appeals to the need to preserve domestic tranquility and family ties.

At its best, bargaining enables negotiators to develop new positions that meet everyone's goals or to develop reciprocal concessions that leave everyone better off. In other situations, bargaining enables the negotiators to avoid destructive conflict through compromise and concession. At its worst, negotiation can heighten animosity and contention.

This book looks at conflict, not between members of a family, but between disputing groups in society. It examines the potential for negotiation and mediation to resolve disputes that arise in a community over the siting and construction of new facilities.

The issues in a dispute vary. Groups can oppose a facility because of its local economic or social impacts, or its effects on local or regional environmental quality, because they prefer an alternative project or because the particular project conflicts with concerns that they hold dear. Although specific issues vary, the disputes invariably involve a project sponsor (either a private firm or a government agency), local groups, and a review process supervised by a government body.

Negotiations are used in these disputes less often than one might expect. Despite the apparent attractiveness of bargaining in many dispute settings, the laws and groups that now shape these conflicts generate obstacles to negotiations. The lack of bargaining can lead opposing groups, often developers and community groups, into bitter litigious struggles that can prove especially costly.

In disputes over proposed development projects, opponents often tie their objections to the environmental impacts that the proposed development project will produce. For this reason, efforts to resolve the dispute between the community and a project developer through bargaining are sometimes called *environmental negotiations*. Many of these disputes will include the participation of a neutral third party who facilitates the negotiation. This subclass of bargaining that attempts to mediate a dispute will be called *environmental mediation*. Cormick, a prominent practitioner and developer of environmental dispute resolution, describes environmental mediation:

> Mediation is a voluntary process in which those involved in a dispute jointly explore and reconcile their differences. The mediator has no authority to impose a settlement. His or her strength lies in the ability to assist the parties in resolving their own differences. The mediated dispute is settled when the parties themselves reach what they consider to be a workable solution. (Cormick, 1980, p. 27)

Thus, in contrast to arbitration (where the third party has the task of settling the differences that separate the parties), the effectiveness of mediators depends not on their authority or wisdom, but on the acceptance of the disputants of their suggestions and aid. This book evaluates these efforts and identifies policy changes that could enhance the chances of resolving disputes over development projects through bargaining.

Although introspection provides a good orientation toward a study of bargaining between individuals, negotiation between disputing groups is better understood through the use of more formal tools of social science analysis. This book uses three techniques to analyze existing and potential processes for facilitating environmental negotiations: (1) analysis of cases of environmental negotiation and mediation; (2) a comparative analysis of the groups and institutions in development conflicts, international relations, and labor–management relations; and (3) a strategic analysis of the incentives acting on bargaining participants.

The study of existing case studies permits one to hypothesize which factors influence the chances that a dispute may be resolved through negotiations. Analyzing the events that occurred in specific instances of negotiation provides a factual grounding for discussion. This anchors discourse to a range that offers practical value and insight. Furthermore, detailed case studies also enable one to see the importance of context for explaining specific events.

Unfortunately, case studies have several inherent methodological weaknesses. Conclusions drawn from case analysis run the risk that they will rest on individual circumstances rather than on factors present in all situations. Further, examining cases limited in time or setting can cause a researcher to overlook larger historical or geographic trends. Nevertheless, the straightforwardness of this technique enables other analysts to dissect the facts for themselves and actively develop competing hypotheses. In addition, if statistical analysis is not possible, as in this emerging policy field, then the analysis of individual cases can help one form hypotheses and reach conclusions. When other analyses support the findings of case studies, then one can place more confidence in the conclusions.

Environmental negotiations have much in common with other systems of bargaining. A comparison of the institutions that structure groups and their relations in environmental negotiations with those in more formal negotiation systems, such as industrial and international relations, provides information concerning the limitations and potential of different negotiation environments. The study of the emergence and development of United States labor–management relations, for example, enables one to see how laws addressed the difficult questions of how to avoid destructive conflict and violence, how to determine who is a legitimate bargaining representative and what is an appropriate issue agenda, and how to enforce agreements. The information existing on labor–management and international relations offers the opportunity to test our hypotheses concerning environmental negotiation on systems rich in detail. This comparative analysis can help to identify alternative policies toward bargaining over development projects and to predict their results.

Finally, an examination of the incentives affecting participants in negotiation can enable one to predict the performance of a negotiation system. In policy analysis, the model of rational choice—that individuals take actions in their self-interest—often proves an effective starting point. This book assumes that people and groups choose to negotiate when the benefits from negotiation exceed the costs. Further, individuals develop rational strategies and tactics in negotiation to maximize net benefits. In organizations, however, a complex dynamic develops between the shared goals of the group and the incentives acting on the leaders. An application of the theories of group cohesion, action, and leadership developed by Wilson (1973), Truman (1951), and Olson (1973) allows one to assess how the structures of a group affect the incentives of the leaders.

Despite the novelty and rarity of negotiations between developers and local groups in environmental disputes, uses of bargaining are surprisingly well documented. Chapter 1 uses a case study of a controversy over the construction of coal-fired electric plants at Colstrip, Montana, to frame the major issues that arise when attempting to resolve disputes through negotiation. Other case studies of environmental negotiations and mediation suggest the potential of this technique as an alternative to existing adversarial forms of dispute resolution.

Chapter 2 applies the model of rational choice to an examination of the most common reasons that negotiations fail. This review of the writings of many bargaining theorists and practitioners adds a note of caution: groups will not reach consensus simply by discussing their differences. Negotiations will always fail to produce a settlement when fundamental differences and interests preclude the existence of a balancing of issues that the bargainers will accept. In addition, many other negotiations fail because the parties, working between themselves, are unable to overcome the institutional, organizational, and communications barriers that often limit the give-and-take necessary for discovering arrangements that all would prefer. Later chapters examine how these common barriers to agreements have a special relevance in development disputes that are contested on environmental grounds. This analysis can serve as a guide to the assessment of the current and future potential of negotiations for resolving these development disputes.

Chapter 3 begins an examination of the structures of labor–management and international relations to distill those elements of these negotiation systems that contribute to the success or failure of bargaining efforts. The groups participating in these negotiations possess a diversity that enables the analyst to predict how the membership structure, the range of member interests, and the bargaining skills of the leadership alter negotiation behavior. In particular, law supports the use of negotiations in labor–management disputes and regulates both membership and leadership structures, while international relations involve groups with leadership that varies in structure and stability. The formal structure of labor unions, where membership is compulsory and leaders are elected to negotiate, offers a very sharp contrast with the structures of many groups in development disputes, where membership is voluntary and charismatic leaders take principled stands.

Chapter 4 turns the analysis from the structure of groups to the structure of the legal and institutional environment in which they interact. Negotiation and mediation always take place within a wider framework of laws, past relations, and institutional settings that control the frequency, the type, and the structure of disputes. Laws and customs in labor–management and international relations help groups to recognize each other, to form negotiation agendas, to create bargaining deadlines, and to structure their bargaining relations. The lack of formal support mechanisms in environmental disputes creates obstacles that participants in ad hoc procedures must overcome. In the absence of these structures, however, mediators can help interested disputants to develop modes of interaction that facilitate negotiation.

Despite their importance, organizational and institutional structures neither create nor resolve a dispute. Within any particular dispute, there exists a structure of issues and power that shapes the strategic and tactical dimensions of the bargaining. Chapter 5 examines how the issue agenda, the preferences of the disputants, and the bargaining tactics used in a dispute setting affect the shape and nature of the final agreement (or disagreement). Since this chapter applies the model of rational choice and the methodology of negotiation theory to disputes, it allows us to identify those tactics most likely to destroy a chance for settlement and suggests how mediators can help disputants to overcome the destructive consequences of particular bargaining tactics.

The process of negotiation requires communication. In any negotiation, bargainers face mixed incentives. Each wishes to advance his or her own interest, yet each must determine whether an agreement exists that offers gains over nonagreement. To determine whether such an agreement is possible, bargainers must communicate with each other and exchange information. Chapter 6 examines how mediators create an additional message channel that negotiators can use with fewer fears that a bargainer will use communications to exploit his or her opponents.

Chapter 7 continues the analysis of the mediator's role in bargaining. It explores how mediators, acting with the consent of the bargaining participants, can use their power to overcome obstacles and facilitate the achievement of negotiated settlements.

For negotiation or mediation to succeed, both bargainers and mediators must possess the ability to use their discretion in developing and adopting bargaining settlements. Unfortunately, the exercise of discretion by groups and individuals raises questions concerning whether the decision process adequately protects the interests of individuals, particularly the interests of those not included in the bargaining. Several writers have identified this problem, and have made suggestions to help ensure that mediators will treat the bargainers impartially (Cormick, 1982) and that mediators will protect wider public interests (Susskind, 1981). Chapter 8 offers suggestions for designing processes that will limit the exercise of discretion to ranges set by law and that are sensitive to the wider concerns of society.

Chapter 9 contains an analysis of the mediation efforts to prevent the Falkland Islands War between Great Britain and Argentina. This chapter suggests that the success of bargaining depends on the ability and courage of leaders to accept the personal risks that arise in negotiation even when the interests of the group or nation diverge from his or her own concerns. This chapter cautions that a good negotiation process and the participation of skilled mediators cannot overcome problems posed by a leadership that cannot make decisions or that makes its short-term interests paramount. I conclude that those proposing the use of negotiations should develop systems that follow the highly structured model of industrial relations, rather than that of international relations, where poor leadership often limits bargaining.

Chapter 10 identifies the major issues in the design of a negotiation process for resolving development disputes. Chapter 11 concludes with a comparison of ad hoc and formal negotiation processes with reliance on traditional administrative and judicial means for resolving development disputes.

Environmental negotiations offer a potentially productive alternative to existing procedures for resolving disputes, but numerous obstacles currently limit its use. I hope that the reader will come to share this view. Perhaps more important, this book will give the reader new insights into how negotiation systems work, and how policy and law can support bargaining. The question of whether to pursue this alternative is fundamentally a political question. Analysis cannot make that choice for an individual, but it can help to focus the debate on the key issues. I hope that the reader will judge that this book does so.

CHAPTER 1

Why Negotiate or Mediate Development Conflicts?

Introduction

In 1972, a consortium of five electric utilities led by the Montana Power Company announced a decision to construct two 700-megawatt coal-burning plants at Colstrip, Montana. These plants would join two plants already generating electricity at the Colstrip site. A combination of low-sulfur Montana coal and sophisticated flue gas desulfurization devices (scrubbers) enabled the two existing plants to rank among the cleanest power plants in the country. The newly proposed plants, although larger, would also burn the same low-sulfur coal and use scrubbers to control emissions. Utility officials expected to receive routine approvals from all state and federal reviewing agencies (Sullivan, 1984).

The expectations of these utility officials proved wrong. The first permit filing in June 1973 was followed by a battle between the power utilities, opposition groups, the Northern Cheyenne Indian tribe, and state and federal regulatory agencies that continued for almost seven years. The battle raged in administrative and judicial proceedings at the state and federal level. This dispute developed into one of the most bitter and lengthy ever witnessed over the construction of a power plant that burns fossil fuels.

Although disputes involving the regulatory officials, Montana Power, and a ranchers' organization helped to prolong this controversy, the conflict between the Montana Power Company and the Northern Cheyenne tribe was in many ways the most bitter and important. The

9

Northern Cheyenne opposed expansion at the Colstrip site for a variety of environmental, economic, and cultural reasons.

The Colstrip Power Plant site lies about 20 miles north of the Northern Cheyenne reservation. Although it is not possible to see the plant from most of the reservation, the construction and operation of the plant would affect the life of the Northern Cheyenne. In the dry plains states of Big Sky Country, clear days allow one to see close to 200 miles. Emissions could dramatically reduce this range. Further, the emissions from the plant would lead to decreases in the ambient air quality over portions of the reservation. Although much uncertainty surrounds the degree of deterioration that would result, most experts agreed that the deterioration would be small. Even with the operation of two more plants, the airshed over the Northern Plains would remain among the cleanest in the nation. Nevertheless, some members of the tribe believed that part of their heritage as Indians lay in the clean air and water of their reservation lands and opposed any project that would lead to environmental deterioration.

Other tribal leaders felt that the construction and operation of the plant would prove disruptive to tribal life. The construction of the plants would cause the population of Colstrip to double from 2,000 to 4,000 individuals. After project completion, the population would return to preconstruction levels. Such rapid expansion and contraction would seriously affect the town of Colstrip, the Northern Cheyenne reservation, and the surrounding county. In other towns of the Northern Plains, such rapid growth has led to increases in crime, to higher expenses for education, to high housing costs, and to incidents between the construction workers and the local population.

In addition to these predictable effects of energy development, some influential members of the Northern Cheyenne saw the construction of this facility as a threat to traditional tribal values. In this view, the increased wealth and the contact with individuals with lives not tied to the land would disrupt the tribe. They feared that the assimilation of young Indians into an industrial workforce would erode the ties that arise from ranching reservation lands.

Montana Power officials felt that they had a legal duty to provide customers with inexpensive power. Further, they believed that power production helped meet the nation's energy needs. In addition, they believed that they had acted as responsible developers in the past, and that this project would continue their good record. Indeed, Montana Power's company town of Colstrip had won awards for the quality of its design (Sullivan, 1984).

Each side spent small fortunes in legal fees promoting their case and challenging the outcome of regulatory decisions on a variety of substantive and procedural grounds. The first forum for contention was the state regulatory agencies and the state courts. The Northern Cheyenne participated formally in the very first regulatory reviews undertaken by Montana's Department of Natural Resources in June 1973. After a series of public hearings and detailed environmental reviews, the Board of Natural Resources and Conservation voted four to three in July 1976 to grant Montana Power a certificate of environmental compatibility and public need. This action was immediately appealed by the Northern Cheyenne and a consortium of ranchers and environmental advocates. In March of 1978, the Montana District Court issued an injunction halting construction. Montana Power appealed this decision. Although Montana Power persuaded the Montana Supreme Court to stay the injunction in April, construction was stopped and remained halted pending final judicial resolution of the major issues in dispute.

Federal legislation created other grounds for contention and maneuver. The federal government passed the Clean Air Act in 1970 and amended it in 1977. This act requires the Environmental Protection Agency (EPA) to set air-pollution control standards that all new power plants must meet and to review all project designs before the start of construction. Furthermore, this act divided the country's airsheds into three classes, specified how much additional pollution an airshed can absorb, and established procedures for redesignating the class of a particular airshed. State governors and Indian tribes were given the statutory authority to request redesignation. Finally, the regulations implementing this act exempted from the regulations plants that had started construction before the development of the regulatory program.

The simple task of determining when construction had started on the Colstrip plants produced a bitter fight between EPA and Montana Power, and it illustrates the type of uncertainty that can plague regulatory programs and judicial decision-making. Despite the start of physical construction at the Colstrip site in 1974, EPA in 1976 ruled that construction had not commenced by June of 1975, the time for exemption from the regulations. This decision rested on a narrow interpretation of the regulations, and Montana Power contended that it contradicted verbal assurances of exemption given to it by an EPA employee. A series of judicial contests followed this decision, sparked as much by a sense of injustice as by corporate interest. In 1977, Montana Power won its appeal of the EPA decision in the Federal District Court. In 1979, however, the Ninth Circuit Court of Appeals reversed this decision.

The need to comply with the new federal regulatory program created room for still more contention and altered the perceived ability of the Northern Cheyenne and Montana Power to achieve their goals. While appealing EPA's determination that new plants must meet all the requirements of the new regulatory program, Montana Power prepared an application to EPA for the required permits. Meanwhile, the Northern Cheyenne prepared a request for a redesignation of the airshed over the reservation lands. This redesignation would subject all new polluting facilities in the region to stricter environmental reviews and could lead to a ban on the construction of new sources of pollution, such as the Colstrip power plants. EPA granted this redesignation in August of 1977.

To meet the stricter standards that followed redesignation, Montana Power amended its initial application to include additional data and a program to monitor air quality. In January 1978, EPA announced its intention to issue a permit to Montana Power. A series of public hearings, however, generated new data that subsequently led EPA to reject Montana Power's permit application. After unsuccessful legal moves, Montana Power submitted still another revised permit application in February of 1979 that offered to use dolomitic lime to enhance the efficiency of the flue gas scrubber. Following still another series of public hearings and discussions, including unsuccessful negotiations with representatives of Montana Power and the Northern Cheyenne tribe, EPA issued a permit in September of 1979. The Northern Cheyenne immediately began legal action to appeal EPA's decision and to prevent the restarting of construction.

This long and tortured battle was, however, almost over. Under the provisions of the Montana State Siting Act, energy developers must take actions to mitigate the adverse consequences of energy development and to compensate communities for local impacts. Despite the bitter fight between Montana Power and the Northern Cheyenne, representatives of each group began a series of negotiations in October 1979 to resolve remaining issues and to develop a plan to address the consequences of construction. In April 1980, after several months of direct negotiation over a broad agenda that included not only environmental but also economic issues, Montana Power and the representatives of the Northern Cheyenne's tribal council reached a comprehensive settlement. The Northern Cheyenne agreed to drop the remaining legal challenges to the plant. Montana Power agreed to establish a special job and scholarship program for the Northern Cheyenne, bus service from the plant site to the construction site, payments to the tribe for increased police protection, payments to sponsor an air-quality monitoring program, and other compensation. Construction restarted in 1980.

This final agreement left both the Indian tribe and Montana Power in a positive frame of mind. Although one may argue that the two sides had merely worn each other out and that the final agreement resulted as much from exhaustion as any other sentiment, the participants did not view it that way. Montana Power officials felt that the agreement would bring it better relations with the community surrounding its project. The tribal representatives believed that the negotiated agreement would enable the Northern Cheyenne to protect their economic, environmental, and cultural interests. All wished that they had adopted a negotiation approach to the dispute earlier and felt that negotiation offered the possibility of a better and cooperative approach to solving development problems.

The events surrounding the Colstrip Power Plant controversy offered both participants and observers a series of surprises. How could a dispute over a conventional power technology using advanced pollution-control equipment continue so long without a final decision, whether for or against the project? How did parties who fought so bitterly with each other finally get together to negotiate their final differences? Why didn't they negotiate sooner, before years had passed and legal action had transformed a concern for environmental and local issues into an arcane discussion of narrow technical issues? Why did negotiations sponsored by EPA fail in September of 1979, but direct bargaining between Montana Power and the Northern Cheyenne succeed in the spring of 1980?

For those concerned with the functioning of government, this case raises a another set of important questions. Is this case atypical of environmental reviews, or a dramatic example of how regulatory traps can subvert even well-planned facilities? What steps could government take to discourage seemingly endless litigation while protecting both national goals (such as environmental quality and the security of energy supplies) and local desires for community autonomy? Does the negotiation of development disputes offer government a practical alternative to adversarial regulatory and judicial procedures?

Environmental Conflicts: Destructive Battles?

In the last decade, the nation's concern for the protection of environmental quality led to the adoption of a broad range of laws and programs. The National Environmental Policy Act (1969) ended the 1960s and ushered in a decade of major federal and state environmental legislation. Congress passed the Clean Air Act, the Clean Water Act, the Toxic

Substances Control Act, the Resources Conservation and Recovery Act, and the Superfund Act. Other acts sought to protect the public from the use of pesticides, from contaminants in drinking water, and from radioactive waste. These laws subjected business operations to public scrutiny and gave interested citizens new powers to participate in the environmental reviews for many new facilities. Judicial interpretation of these laws has further expanded the rights of citizens to voice their concerns (Susskind and Weinstein, 1980).

By the decade's end, citizens and any interested group possessed an ability to affect regulatory decision-making through their participation in environmental reviews (Stewart, 1975). Nevertheless, the institutions used to incorporate this participation often channeled disputes over specific projects into particularly destructive paths. Public hearings and formal intervention at regulatory proceedings produced many public debates characterized less by a full discussion of the issues than by strategic litigation and delay.[1] With final decision-making resting in the regulatory agency or court and issues circumscribed by technical concerns, each side in a dispute faced incentives to portray the adversary's goals and proposals in the worst possible light and frame their objections in legal terms. Cooperation or collaborative planning could jeopardize a legal case. The development process became one in which developers decided where to build, announced their decision, and then defended it in a series of regulatory and judicial trials (Ducsik, 1978).

Conflicts between proponents and opponents of development grew to be protracted and expensive (Friedman, 1979). Judicial and administrative contests that resemble the darkest chapters of Dickens's *Bleak House* have crowded the pages of books and newspapers. In strongly contested proceedings, the New England Power Company spent $30 million on design, licensing, and litigation for a proposed nuclear power plant at Charlestown, Rhode Island ("New England Power . . .," 1979). Subsequently, the utility abandoned the project in the face of rising public opposition and growing disenchantment with nuclear power, rising interest rates, and decreased demand for electricity. Oil refineries in New England have met with similar fates. Despite statements of each of the New England governors supporting the siting of a regional oil refinery, nine attempts have failed to provide New England with its first (O'Hare, 1977). In California, proposed housing projects have run afoul of citizen opposition and judicial decision-making. In one dispute, a Boy Scout, seeking a merit badge, intervened against a 200-unit condominium

---

[1]See Frieden (1979, p. 65) for a particularly dramatic example of how litigation leads to paralysis.

project. Although the project was eventually approved, the developer estimated that legal fees and other costs totaled over $400,000 (Frieden, 1979). The ability to force developers to pass their proposals through complex regulatory decisions has helped opposing groups and individuals to contest and stop development projects.

Although these individual examples show large costs, when projects are expensive, the costs of delay and detailed reviews will generally amount to a small percentage of what the entire project costs. This is especially true as long as the delays and reviews occur in the preconstruction phase of the project. Once construction starts, however, the developer will incur heavy losses if a project is halted. Finance charges mount, and labor costs also rise as work crews are disbanded and later reassembled to complete projects. Furthermore, a judicial reversal of permits can jeopardize an investment. Thus, the uncertainties that complex regulatory procedures can introduce into a project are more important than review costs.

The Colstrip Power Plant controversy illustrates the heavy economic risks of regulatory uncertainty. In the Colstrip power controversy, Montana Power officials began their initial construction believing that they were not subject to the provisions of the Clean Air Act. A court case resolving the ambiguities in the statutory language of the Clean Air Act left Montana Power, after a $200-million investment, subject to regulatory procedures that could have forbidden further construction. The lack of a permit led to a costly halt in construction of the Colstrip site (Sullivan, 1984).

Uncertainty often arises from the large number of permits and reviews required for many projects. These add to the complexity of the review process and can make the final outcome uncertain. Unfortunately, to the degree that a process possesses uncertainty, it will prove impossible for the developers to take actions that would screen out unacceptable projects before committing resources to detailed design studies and permit applications. For an unsuccessful project, these resources become sunk costs, lost to the developer, but fundamentally altering the cost of producing new facilities. If some projects fail to gain an approval after a heavy investment, then those projects which are completed must carry the costs spent in the planning and design studies of those disapproved.

The financial risks imposed by regulatory uncertainty provide few environmental or planning benefits. It is the expenditures on pollution abatement equipment, design modifications, or the reasoned denial of construction permits that produce the environmental protection envisioned in the laws. The uncertainty that a developer faces when considering a specific project produces few benefits. If procedures were clear and easy to understand, developers would propose and develop only those projects that would meet the standards set in the environmental regulations.

The search for crystal-clear procedures, although helpful, will neither resolve all uncertainties nor remove conflict. Environmental laws generally require that administrators apply criteria to a set of diverse situations. A developer must determine how to characterize the proposed project and indicate how it meets the relevant standards. The government agency must then consider the evidence that the developer presents to determine which standards a project must meet and whether it does. These determinations can differ. In addition, local conditions and concerns affect a siting review process and lead regulators to exercise further discretion in making decisions.

The specific reasons that development projects provoke disputes are as diverse as the projects and the communities in which they are located. One can, however, identify four major sources of conflict[2]:

1. Disagreement over the relevant weights granted to competing policies and values
2. Disagreements over the the new distribution of costs and benefits that arise from a project
3. Disagreements over the appropriate level of protection from environmental and health harms
4. Disagreements over the use of fixed resources

Often, several of these sources contribute to a particular controversy. In the Colstrip controversy, the Northern Cheyenne fought Montana Power because they perceived that the development project would threaten a way of life that they valued over economic development. Montana Power, on the other hand, placed great weight on its policy objective of providing low-cost power to Montana residents. The Northern Cheyenne also opposed the project because it would impose costs on the tribe for the needed police and housing services; Montana Power viewed Colstrip as an excellent site for low-cost energy production. Finally, the EPA and Montana Power fought over the specific pollution-control measures needed to protect the air quality. Thus this one dispute involved conflicts over values, the distribution of costs and benefits, and the standards for pollution control.

Disputes over values, economic impacts, standards, and the use of fixed resources often produce an administrative battle followed by an adversarial judicial contest. The Colstrip controversy is typical: the dis-

---

[2]This scheme is based on the work of Susskind, Richardson, and Hildebrand (1978).

putants contested many issues in administrative forums, and the losers appealed to the courts. In the Colstrip dispute, the repeated reversal of lower court decisions suggests the difficulty that jurists have in resolving these troubling conflicts and also suggests the uncertainties disputants face concerning their rights and powers. Furthermore, in the Colstrip controversy, regulatory and judicial proceedings necessarily focused on narrow issues that were recognized in federal law—the setting of a regulatory standard—rather than the broader concerns that affected the community. Thus the disputants initially debated environmental issues rather than the community issues that were also principal concerns.

Dissatisfaction with the actions of regulatory agencies and the courts in resolving these controversies has proved a common theme in recent political and legal analysis. Numerous economic, political, and organizational theories argued that government decisions reflected the influence of special-interest groups rather than the application of scientific techniques that would promote the public welfare (Kolko, 1963; Stigler, 1971). Theories of "regulatory capture" argued that government agencies acted to serve rather than control the industries they were empowered to regulate or evolved toward senility and inaction (Bernstein, 1955). Expertise itself has faced a general attack, and many have come to realize that technical analysis requires analysts to make value judgments (Bacow, 1982; Tribe, 1976). Stewart (1975) traces the evolution of judicial theories used to resolve these conflicts and argues that current administrative law lacks a principle that can adequately guide jurists in resolving these thorny political problems. Others question whether one can reasonably expect courts to resolve these conflicts (Rifkind, 1976). This dissatisfaction has arisen at the same time as an increase in the number of disputes requiring judicial resolution (Barton, 1975).

Negotiated Resolution of Environmental Disputes

In the Colstrip Power Plant dispute, after years of litigation, representatives of the Montana Power Company and the Northern Cheyenne tribe sat down and negotiated a resolution of the issues that separated them. Unlike the judicial and regulatory procedures in the early days of the dispute, these discussions addressed the economic, environmental, and social implications of the construction activity. Working over a winter, the representatives developed an agreement that included measures to mitigate adverse impacts of the construction and to compensate the Northern Cheyenne for some of the expenses that they would bear. A discussion of a broad range of issues concerning the future of the relationship between the Colstrip community, Montana Power, and the Northern Cheyenne enabled both sides to swiftly conclude the dispute that each had participated in for such a long period. Individuals on both sides felt that they should have attempted to negotiate their dispute sooner, that their negotiation had achieved positive results, and that they parted with positive views of each other (Sullivan, 1984).

Negotiation became possible for several reasons. The Montana Siting Act legitimated a wide-ranging discussion between the utility and the local community. Thus the long legal battles had helped each side to form a clearer view of the new power relation between them. The Northern Cheyenne realized that they did not have the power to stop the facility and Montana Power realized the political and economic importance of community assent. EPA's permit decision had resolved the issue of determining the appropriate level of pollution control. Working together, Montana Power and the Northern Cheyenne were able to develop an agreement that mitigated other adverse impacts and compensated the tribe for economic costs. Furthermore, the support of cultural exchanges between the tribe and the workers offered a settlement whose framework embraced the diverse values and concerns of participants.

This outcome, although arising only after years of litigation and confrontation, has occurred elsewhere without such a destructive struggle. Consider the following examples of negotiated dispute resolution.

Snoqualmie River Dispute

A proposed dam on the Middle Fork of the Snoqualmie River in Washington State formed the center of this dispute, which was among the first development disputes to use formal mediation (Cormick and McCarthy, 1974). After a 1959 flood, residents living along the river's banks and downstream farmers desired flood control measures. The local county sponsored an Army Corps of Engineers study that led to a dam proposal (Mazmanian and Nienaber, 1979). Opponents of the project feared that a dam would spawn suburban development of the flood plain by Seattle residents, and they objected to the loss of the free-flowing river.

The Corps requires the approval of the governor of a state before it seeks funding for any project. Governor Evans of Washington opposed the dam proposal in 1972 but wished to take some steps to alleviate the problem of flooding. Governor Evans appointed leaders of the Environmental Mediation Project of the University of Washington as mediators in 1974.

The mediators initiated discussions with a group of 10 individuals with influence and stature in the conflicting groups. Meetings included a member of the Corps who could sketch alternatives to the current project. Through discussions, opponents of the dam discovered that the farmers did not desire to sell their land for development, and that they would agree to support zoning restrictions to limit flood plain development. Dam proponents established that flooding caused them economic hardship by destroying their crops, and that continued flooding would not provide an acceptable solution. They made dam opponents believe that they would be held politically responsible for any damages from a future flood. They stressed that a flood would destroy the regional credibility of environmentalists and lead to the construction of a dam without any amenities or land use restrictions.

A negotiated agreement called for a dam on the North Fork (rather than on the disputed Middle Fork) of the Snoqualmie River, a series of levees and setbacks along the Middle Fork, and land use and zoning restrictions on the downstream farmland. A basin planning council was established to plan growth, and the state agreed to buy development rights in the flood plain.

White Flint Shopping Mall

In White Flint, Maryland, representatives of a department store chain met with local community residents to develop a negotiated approach to the siting and design of a shopping mall. With the assistance of a mediator, they concluded negotiations with an agreement that called for several design restrictions to limit the impact of the shopping center on the local community. The department store also agreed to construct a landscaped slope separating the shopping center from the neighborhood. Further, the department store agreed to compensate local residents who wished to sell their property for any decrease in property values that occurred in the five years following project completion (Rivkin, 1977).

Brayton Point Coal Conversion

In Massachusetts, representatives of New England Power Company (NEPCO), the U.S. Department of Energy, the U.S. Environmental Protection Agency, the Massachusetts Department of Environmental Quality Engineering (DEQE) met under the auspices of the Center for Energy Policy to consider the conversion to coal of an oil-burning power plant at Brayton Point. An 11-month mediation process led to an agreement in which NEPCO agreed to install additional pollution-control equipment and to use low-sulfur coal in return for the adoption of a new regulation by the DEQE that would set pollution limitations for 10 years. This mediation process enabled agencies to coordinate their different regulatory reviews and to develop a consistent interpretation of the implications of energy and environmental laws (Burgess and Smith, 1984).

These successful efforts to resolve environmental disputes through negotiation allow us to describe this process. As now practiced, environmental negotiation consists of voluntary meetings between representatives of the disputing groups. The effort to resolve the disputes generally proceeds in parallel with formal regulatory and judicial reviews of the proposed project. Negotiation procedures and agendas are determined by the participants, and agreements often have an unclear legal status. When these negotiation efforts involve a third-party intervenor, they are commonly called *environmental mediation*. Although the activities of the mediator can range from simply procedural (such as running meetings) to more substantive (suggesting issues and settlements), the mediator lacks the ability to decide the issues. A settlement is binding only when the parties assent to it (Cormick and Patton, 1977). The lack of formal structure in these disputes has frequently led to the participation by mediators in environmental negotiations.

The fortunate outcomes of these and other bargaining efforts suggest that the potential of negotiations in resolving these disputes is great. Nevertheless, in comparing negotiation efforts with traditional processes for resolving disputes, one must develop consistent criteria that permit one to determine both the advantages and the disadvantages of each. Developing criteria for evaluating a negotiation effort is quite difficult. Is a negotiation effort a failure because no agreement is reached? Is it a success if an agreement is reached after a long and difficult conflict? Fisher (1979) offers perhaps the best answer in his concept of principled negotiation. Principled negotiation rests on the assumption that

> the proper standards for judging a process of conflict resolution are *not* those that may produce a particular result in a particular case, but rather those standards that will tend to produce desired results in an indefinite series of cases. (p. 1)

He develops criteria for assessing a negotiation process. These can help us to assess the success of efforts to resolve environmental disputes. Slightly modified to facilitate a comparison of negotiations with other forms of dispute resolution and changed in order of presentation, they are:

1. Do the process and outcome of a dispute resolution effort appear fair?
2. Is the decision or settlement consistent with preexisting practice?
3. Are the results acceptable to the parties?
4. Does the process improve relations between the parties?
5. Does the process produce quick, low-cost decisions?
6. Does the process reconcile the interests of the parties in ways that leave no opportunities for mutual gain?
7. Does the decision or settlement set a good precedent for the future?

These criteria may be grouped into those that concern the fairness of the dispute resolution mechanism (1–4) and those that concern its efficiency (5–7). These criteria will enable us both to compare processes for resolving disputes and to guide the design of alternative processes.

Through comparison, we can see both the promise and the risks that negotiation processes hold. Negotiations may offer a process that is fairer than adversarial procedures, but traditional adjudicatory procedures possess the legitimacy of law and custom. Negotiations will never have the concern for precedent that dominates the traditional process. Negotiations, however, should facilitate an improvement in relations between the parties. Although negotiations may produce outcomes that are fairer and more efficient than traditional procedures, only experience allows one to know this with certainty.

Perspectives on Environmental Negotiation and Mediation

Despite the newness of the application of negotiations to environmental disputes, there already exists quite an extensive literature on the subject. This literature includes number of case histories of environmental disputes.[3] The earliest analytic writings on the subject both define the process of environmental mediation and draw conclusions from the experience of practitioners (Cormick and McCarthy, 1974; Cormick and Patton, 1977). These first analytic efforts grew out of the community conflict and public participation fields and focus on the timing of an intervention and the techniques of mediation. These writings express a particular concern for avoiding premature intervention by third parties in a

---

[3]Talbot (1983) offers a collection of six case studies of environmental mediation. Susskind (1981) provides a list of disputes settled through negotiation or mediation.

dispute because there is often an unclear power relationship between the formal party and the less organized community. Unfortunately, such an approach can miss many opportunities for avoiding destructive conflict. Stalemates commonly occur only when competing groups have reached an impasse in adversarial procedures.

Other early approaches stress the need for structuring the negotiation process and see the mediator as a source of that structure. Susskind, Richardson, and Hildebrand (1978) (and also in Susskind and Weinstein, 1980) identify steps for resolving ad hoc environmental disputes. These steps include:

1. Identifying the parties that have a stake in the outcome;
2. Ensuring that each interest group is adequately represented;
3. Identifying the key issues and narrowing the agenda of points in conflict;
4. Generating a sufficient number of alternatives;
5. Agreeing on the boundaries and time horizon for impact assessment;
6. Weighting, scaling, and amalgamating judgments about costs and benefits;
7. Determining fair compensation and possible compensatory actions;
8. Implementing the bargains that are made; and
9. Holding the parties to their commitments.

This study provides a practical guide for those involved in dispute resolution and offers a vision of negotiation as a way of constructively fostering community participation in development decisions. It identifies the problems associated with the lack of structure that often complicate these dispute settings, and it offers a way for an intervenor to supply a dispute with the structures necessary to make the give-and-take of good-faith bargaining possible.

Scott Mernitz (1980) in *Mediation of Environmental Disputes: A Sourcebook* both examines the emergence of environmental conflicts and identifies different roles that mediators can play in resolving them constructively. This book seeks to serve as a handbook for mediators. He identifies the most important attributes for determining the suitability of disputes for mediation:

(1) stage of conflict; whether an impasse has been reached;
(2) parties to dispute; their sizes and sanctions;
(3) geographic scope of the dispute;
(4) role of the state governor in the conflict;
(5) negotiable, ancillary factors that could produce settlement;

(6)   social and psychological factors, notably facesaving, communicative processes, and interaction during negotiation, which the mediator could only evaluate first-hand; and

(7)   means for implementation of the agreement. (p. 120)

This text also provides a review of case studies of environmental conflicts and lists methods and guidelines for implementing a successful mediation effort.

All these writers have recognized that labor–management bargaining has used mediation quite extensively, and they have used the labor literature especially to suggest mediation techniques (Cormick and Patton, 1977; Mernitz, 1980). More recent structural analyses have attempted to view environmental conflict resolution through negotiations as a setting for bargaining that shares many important dimensions with industrial relations but possesses substantial differences (Sullivan, 1980; Susskind and Weinstein, 1980). Susskind and Weinstein (1980) note that environmental disputes differ from labor–management disputes in several ways that, in their view, merit special attention:

Irreversible ecological effects may be involved; the nature, boundaries, participants, and costs are often indeterminate; one or more of the parties to most environmental disputes often [claim] to represent the broader public interest (including the interests of inanimate objects, wildlife, and generations yet unborn); and implementation of private agreements is difficult. (p. 324)

Once one recognizes that we are discussing not a unique phenomenon but a system of controlling conflicts between competing groups, then the links with labor–management and international relations offer fertile ground for exploring the potential of this technique. Several classics offer insights particularly relevant. Ullman's (1955) history of the development of the trade union movement offers an exceptional analysis of how the system of labor–management relations developed from ad hoc forms of interactions to one that is sharply controlled by law, regulation, and tradition. Dunlop (1958) analyzes how different nations have adopted different structures in their labor relations and enables one to see that no single course of development was inevitable. Simkin (1971), a former head of the Federal Mediation and Conciliation Service, provides an insider's view of the nature of mediator actions, the rules of thumb used to guide interventions, and the relationships between the mediation service and its clients. Ann Douglas (1962) documents particular cases of bargaining and mediation. Her work enables one to see the technical actions of bargaining participants in specific negotiations.

International relations offers another model of negotiation that enables one to see how bargaining takes place when no higher authority controls conflict. Once again, a wide literature that contrasts the diplomatic styles and operations of nations (Kissinger, 1977; Iklé, 1964) enables one to see how systems of domestic structure and leadership accountability affect how bargaining develops. Others (Jackson, 1952; Fisher, 1969; Young, 1970) have analyzed the actions of mediators in particular international disputes to show how these actors overcame the problems of enforceability of agreements and internal political problems, as well as the problems that arose from widely disparate views of the world.

Negotiation has spawned a literature of its own that deals with strategy and tactics. The strategic analysis of negotiation, both abstractly (Schelling, 1960; Raiffa, 1982; Fisher and Ury, 1981; Cohen, 1980) and within an institutional setting (Stevens, 1963; Iklé, 1964; Young, 1970; Kissinger, 1977), suggests the kinds of tactics and problems that any type of negotiation process can face. This analysis helps us to view the mediator as a strategic actor in bargaining and distinguish the contributions that one can make.

Psychologists have studied negotiation as a technique for managing conflict (Wall, 1981; Deutsch, 1973; Rubin and Brown, 1975) and examined how mediators can function in managing the communications between the negotiating parties (Pruitt, 1981). Wall (1981), in an excellent survey article, outlines the specific tactics that a mediator can take and discusses the links between negotiators, mediators, constituents, and external publics.

Finally, lawyers have also examined systems of bargaining and mediation as alternatives to litigation. Fuller (1971, 1978) discusses the limits of current adjudicatory proceedings for resolving disputes involving several viewpoints and argues that some forms of mediation or arbitration may prove more productive. Barton (1975) discusses the explosion of litigation. Sander (1976) outlines alternatives to adversarial judicial proceedings and develops criteria for determining which process best suits a particular dispute.

Conclusion

This book explores the potential and limitations of negotiations and mediation for resolving disputes over development projects. Although each cited case of a negotiated resolution to an environmental dispute was unique, the disputants faced common problems. The disputing groups had to select individuals to represent their interests in the dispute, and each bargainer had to develop skills for effective bargaining.

The bargainers had to develop formal agreements that committed the dispute participants to settlement terms, and each agreement had to survive both judicial and public scrutiny. The resolution of disputes through negotiation becomes possible only when the individual participants surmount these and other obstacles. An understanding of these common obstacles can enable individuals to understand how a negotiation process works, and what law or policy steps can be taken to enhance other efforts to resolve disputes through negotiation.

In my view, the results of ad hoc efforts to resolve development disputes encourage a wider use of negotiations either as a complement to the existing structures for resolving conflicts, or as an alternative path that citizens and government officials may find attractive for certain classes of disputes. The existing case studies of environmental negotiations and the comprehensive institutional and theoretical studies of other systems of bargaining provide organizing concepts that enable one to productively assess the potential of this technique in serving as a supplement or alternative to traditional procedures. A formal analysis of environmental negotiation can enable policymakers to identify what elements are critical to such a system, what limitations it will have, and how well it may work. This analysis should help those attempting to initiate a voluntary effort to resolve a development dispute through negotiations or those seeking to implement a more formal process.

CHAPTER 2

Negotiations Can Fail

Introduction

Homer's *Iliad* begins with sacrilege, injustice, and great anger. King Agamemnon refuses the ransom offered by the priest Chryses for his daughter Chryseis. Chryses prays to the god Apollo, who brings a plague upon the Greek forces seizing Troy. Calchas, a Greek diviner, rightly blames the plague on Agamemnon, and public debate forces him to return his battle prize, Chryseis, to her father.

Agamemnon, lacking a prize, unjustly expropriates Briseis, the battle reward and love of Achilles. Achilles' anger knows no bounds, and only the intervention of the goddess Athena stills his sword. The persuasive words of Nestor, an old and respected leader, fail to defuse the confrontation. Threats escalate, and Achilles refuses to support Agamemnon in battle. Numerous attempts fail to reconcile these two leaders. No offer, no compensation, no honor can assuage the anger of Achilles. Only the death of Patroclus, friend of Achilles, and the threatened collapse of the Greek army can induce Achilles to channel his rage into the battles against Troy.

Old stories contain old truths. Negotiations and mediation cannot resolve all disputes. At times, negotiations serve only to inflame passions and hatred. Agamemnon's threat to take Briseis produced only counter-threats from Achilles and an ever-escalating level of hostile rhetoric. The intervention of Athena and Nestor managed to check Achilles' anger short of physical blows, but the words of these mediators could not settle this dispute. The gross injustice, the great anger, the strong commitments, and the irrationality of the actors that characterized this conflict precluded negotiations. Only the passage of time and new events enabled Achilles to enter battle with honor.

Before one advocates negotiations as an alternative process for making decisions on development projects, one must recognize that negotiations are not a panacea for resolving all disputes. In some conflicts, the disputants will lack the rationality that permits discussion or compromise. In others, disputants may rationally calculate that negotiations offer no advantage over current adversarial forums for resolving a dispute. In others, no agreement will exist. In still others, combinations of factors such as communication barriers, institutional impediments, or organizational ineptitude may prevent disputants from finding a settlement even when the disputants are willing to bargain and could find many resolutions to their dispute.

In disputes over a development proposal, as in any setting, negotiations can offer disputants an opportunity to determine whether a settlement is possible that leaves all or many of the disputants better off while leaving few or none worse off. Unlike adversarial, administrative, or judicial proceedings, negotiations can enable disputants to directly address issues that are a major concern to them and find new arrangements not envisioned in narrow legislation. Negotiation agendas are not limited to issues defined by a law that seeks to protect a particular interest. By linking many issues together into a negotiation agenda, negotiations can enable disputants to determine whether some combination of compromises, concessions, or alternative arrangements can leave all better off. Furthermore, negotiations can enable individuals to avoid surrogate disputes in which they initiate litigation on the issue that offers them the strongest legal case, rather than on the issues that truly trouble them.

Despite these advantages of negotiations for resolving development disputes, the potential of negotiations for resolving a particular dispute will depend on the specifics of the setting, issues, and disputants. When groups believe that they are locked in an epic or heroic struggle, then bargaining cannot succeed. In addition, there are many other potential factors that contribute to the failure of negotiation attempts. Some conditions arise from the structure of a dispute and its setting, but many others arise from technical or tactical bargaining errors.

This chapter develops a list of reasons why negotiations can fail to resolve a dispute. It identifies those conditions that preclude the start of negotiations and discusses how issues and preferences can prevent settlements. Unfortunately, even when prior assessment of the views and positions of the disputing parties would lead all to expect that they could easily reach an agreement, failures will occur. These failures can occur when institutional or organizational structures do not support bargaining, or when bargaining tactics and communication breakdowns subvert a search for agreement. This chapter analyzes how such situations occur and discusses which steps can help disputants reach agreements.

Inability of Negotiations to Start

In disputes over development projects, successful negotiations cannot start unless some disputants possess both a solid rationality and a willingness to explore different ways of reconciling their opposing interests. This exploration necessitates that groups and individuals understand viewpoints and interests different from their own. Although at times some new arrangement can satisfy everyone's interests, negotiations often require that bargainers discuss compromises and alternative offers. When groups are unwilling or unable to engage in discussion or to entertain other views, a negotiation process will pose a threat to their organizational norms and ideological beliefs. For these groups and their leaders, adversarial and judicial contests can transfer responsibility for decision from them to others. In particular, a spirited legal case can exonerate a group from responsibility for the final outcome that results from a judge's decision. Courtroom procedures can enable disputants to avoid the political dialogue that negotiations would necessitate and can offer the possibility of winning new powers and rights. When dialogue, new ideas, or compromise threatens an organization or its leaders, negotiations will fail to start.

In some disputes, the struggle is more important than the ultimate outcome. This can be particularly true when a group sees a particular dispute as an opportunity for mobilizing public support or for increasing the power, strength, or membership of its organization (Alinsky, 1971). When individuals or groups value a struggle for itself or for its by-products, then negotiations or other techniques meant to end a dispute will not offer attractive alternatives to the combatants. If groups desire to delay or discourage a development project, or if they hope to insert great uncertainties into a project's future, then court challenges may offer a better dispute forum than negotiations. The uncertainties and delays posed by judicial challenges can discourage investors and can cause those proposing a project to face higher financing costs. To the extent that a disputant desires to create an environment fraught with uncertainty and risk, that disputant will seek forums where norms discourage good-faith bargaining and the search for compromise.

Great passions and principles, whether noble or irrational, can prevent negotiations. Neither Achilles or Agamemnon would discuss the demands or threats tossed between them. Agamemnon would not retreat from seizing Briseis, and Achilles would not accept any compensation for this injustice. When they both became angry, neither could admit that the interests and lives of both were tied together by the cooperation that the battle with Troy demanded.

Conflicts over the siting of large facilities or development projects can include issues and concerns that preclude negotiation or compromise. Nuclear power plants, for example, generate opposition from those who believe that the use of nuclear energy poses long-term threats to the earth's environment. Their concern for the contamination of the earth does not permit compromise. Remote siting of plants, the incorporation of new safety devices, or a solution to long-term disposal problems would not offer an acceptable compromise to those groups, such as the Clamshell Alliance, that oppose the construction of any nuclear power plant per se (Douglas and Wildavsky, 1982).

Finally, negotiations may fail to start when disputants have strengths and weaknesses that vary with a forum. Negotiations often require that two sides take a positive step to resolve their differences. If unilateral action can force a dispute into a forum where one side has a special advantage, then it is unlikely that negotiations will occur at all. In disputes over the environmental impacts of development projects, litigation by environmental groups has won them many successes. Success brings success, and these groups have developed considerable expertise and have helped develop a body of law that affords them opportunities and strength in litigation. Generous judicial interpretations of standing combine with statutory encouragement of citizen suits to enhance the power of environmental groups to delay projects (Friedman, 1979). These laws that facilitate the initiation of legal challenges to development projects complicate the task of those seeking a negotiated resolution of a dispute.

Agreement Precluded by Preferences and Issues

The most fundamental reason why negotiations fail to produce an agreement is that at times the preferences of the disputing groups or individuals are so opposed that no alternative formulation of the problem or combination of compromises and concessions on the current range of issues can leave both sides better off than a resort to more adversarial forms of dispute resolution.

Perhaps more commonly, it is the initial positions and statements of groups that prove an obstacle to bargaining, rather than some profound difference of interests. Here, various negotiation strategies (Fisher and

Ury, 1981) can enable a skillful bargainer to shift discussions from initial positions to the underlying principles and interests. Such an exploration can enable the disputants to develop positions that offer mutual gains (Cohen, 1980). Despite the cause for optimism engendered by these situations, some disputes involve parties whose basic interests are so incompatible that no agreement is possible until the interests or issues change. Discussion may produce only further evidence of the inability to agree.

This failure of preferences and issues to permit an agreement can occur for the many reasons that cause individuals and groups to define particular goals and to select particular issues for attention. Agreement can become possible only through change in the disputing groups or in the dispute itself. New information or new events may cause the group to change their views and preferences. As time passes, the negotiation agenda may expand to include other issues that permit the development of comprehensive settlements. Events may remove the most troublesome issues. Without these changes, no settlement is possible.

At times, discovering whether a settlement is possible is one of the goals of negotiation. In a commercial setting, individuals and firms seek to determine whether tastes, needs, incomes, or preferences permit them to discover a price for which they would exchange goods or services. Even when negotiations fail to produce an agreement, the negotiations enable the participants to discover the prevailing terms in a market. A buyer's failure to reach an agreement with the supplier indicates that the buyer must either raise his or her offer or spend more time searching for someone willing to provide the goods and services. The seller learns that this individual will not meet the seller's prices, and unless others will, the seller may need to cut the price to remain competitive.

Unfortunately, disputes over development projects rarely allow one to simply walk away when underlying interests are incompatible. Large development projects can affect a whole community, and a consensus of interests can prove difficult to find. The failure of preferences and issues to support a consensus often leads to destructive contests. These contests can be resolved only when a group prevails, when the adversarial contest itself alters the preferences and goals of the conflicting groups, or when government imposes a settlement.

In the trilateral dispute between the Montana Power Company, the Northern Cheyenne Indians, and the EPA over the siting of new power plants at Colstrip, Montana, the initial preferences of these groups precluded any negotiated settlement. Each side believed that its position would triumph in the regulatory and judicial proceedings. Only after years of costly lawyers' fees and inconclusive contests did Montana Power and the Northern Cheyenne alter their attitude in ways that permitted them to negotiate a settlement. (EPA, however, won judicial approval of all its major decisions.) As time passed, both Montana Power and the Northern Cheyenne came to believe that unilateral action could not succeed. In addition, EPA's decision on the environmental issues removed a source of conflict from the dispute. As negotiations shifted to proceedings required by the Montana Facility Siting Act, the issue agenda expanded to include the socioeconomic impacts of the dispute. This made a more comprehensive settlement possible. Only when time and events had resolved many uncertainties and provided new information could productive bargaining begin (Sullivan, 1984).

Labor–management relations offer an example of group decision-making in which it is usually clear from the outset that unilateral action cannot succeed. Nevertheless, uncertainties concerning the group's strengths and desires still exist. In labor–management negotiations, management can face uncertainty over the strength of union demands and its willingness to strike. Labor can face uncertainty over the issues important to management and its willingness to take a prolonged strike. These uncertainties affect bargaining, and either miscalculations or strategy can lead to the mounting or provoking of a strike. Unlike in other forms of conflict, rarely can one side win all their goals through a strike. A strike, however, can clearly indicate the importance of an issue to each group. As strike actions resolve uncertainties, bargaining can adapt to new realities. The short duration of most strikes (Simkin, 1971) shows how quickly the structure of collective bargaining can channel destructive forms of conflict—during a strike, both sides lose—into the more constructive activity of determining new working arrangements and levels of compensation. Only rarely do strikes continue for long periods.

Even in Homer's world, events and new realities often held the key to resolving a dispute. Although time tempered the anger of Achilles, only radical changes altered his attitudes toward helping the Greek forces. When the Trojans pushed the Greeks to the sea and set fire to their ships, Achilles let Patroclus lead the Danaän forces into battle. Although Patroclus altered the tide of battle, Hector, leader of the Trojan troops, slew him. Only then did Achilles desire to enter the battle—to avenge the loss of his dearest friend. This change of heart induced Achilles to settle his grudge with Agamemnon. Agammenon responded to this new situation and returned Briseis and the battle treasures to Achilles.

When preferences and interests prohibit the possibility of agreement, then only time, events, and information can alter a group's preferences and the issues in ways that permit a settlement. In environmental negotiations, the failure to reach a settlement usually brings not the catastrophic costs associated with labor strikes, but the slow pace and irksome costs of litigation. Strikes make salient to both parties their common interest in finding an agreement. Preparing for litigation seldom serves as a spark to altering one's view.

Failure of Dispute Structures to Support Bargaining

Bargaining efforts, despite the best intentions of the disputants and the availability of many potential attractive agreements, may fail when existing organization and institutional structures do not support a negotiation effort. Much of this book will detail how different organizational structures can affect the ability of a group to bargain effectively, and how legal structures can affect the success of problem-solving efforts. Without special efforts to channel disputes into forums that facilitate a search for underlying common interests or that promote the give-and-take of healthy compromise, only good fortune can permit negotiations to succeed.

Productive negotiations will not occur unless the disputing organizations and groups possess some structure that permits them to bargain. To bargain, a group must have a leadership that is capable of articulating its interests and of responding to the interests of other competing groups. To succeed, some participants in bargaining must be capable of translating the abstract goals of a group into concrete bargaining proposals. To conclude, bargaining participants must be able either to accept a proposal as a settlement or to abandon negotiations for more adversarial forums.

These necessary bargaining actions require skillful leaders and strong organizational structures. Bargainers must articulate the concerns of a group as a bargaining position and, perhaps most important, help the group to consider and adopt new perspectives on the issues. Diffuse sentiments of constituents must congeal into specific bargaining proposals.

When bargainers cannot find a new alternative that all can immediately accept, a compromise of initial positions may prove essential to a settlement. Compromise necessitates the strongest leadership and organization structures. The give-and-take across the bargaining table also requires a give-and-take within the bargaining group (McKersie, Perry, and Walton, 1965). The internal give-and-take can place severe stress on an organization. If the initial position offered something to every faction within a group, then concessions can require that a group balance the competing concerns of members without causing the group to dissolve.

The stability of a group under the stress of bargaining can vary greatly, both with the structure and character of leadership and with the structure of the organization itself. Groups that possess a strong and stable leadership are best able to accommodate the internal stresses of bargaining. In the face of the New York City budget crises that began in 1975, the Municipal Workers Union, led by its long-time and powerful leader, Victor Gotbaum, was perhaps best able to accommodate the concessions needed to keep the city solvent.

Negotiating with the Wrong Individuals or Groups

Bargaining requires more than an articulation of interests, the formulation of alternatives that permit joint gains, and the making of concessions. The settlement of a dispute requires that those signing an agreement possess the power to implement it. If the leaders of a group are unable to commit the membership to the negotiated actions, then the leaders of the group have little to offer the other participants in a dispute. Although this observation should appear obvious, it is remarkable how often in complex negotiations, particularly those between groups of different cultures, negotiations fail because the wrong individuals negotiate. At times, those bargaining may lack either the authority to make concessions or the power to deliver on their promises.

In the negotiations that followed the seizure of American diplomats by Iranians, initial negotiations between the United States and Iran proved futile. All negotiations failed until the United States came to realize that Iranian government officials lacked the power to negotiate a settlement or to execute a release of the hostages. Only the endorsement of the religious leader Khomeini would induce the captors to release their hostages.

Likewise, in the dispute between Montana Power and the Cheyenne Indians, the initial negotiations sponsored by the EPA brought together the wrong individuals. As part of its permit-review procedure, EPA tried to build a consensus between Montana Power and the Northern Cheyenne on the conditions of the permit. Technicians and lawyers who had helped the Cheyenne make their case opposing the new plants represented the Indians at the negotiation table. Those representatives of the Northern Cheyenne were unprepared to bargain. Since they anticipated that this negotiation session would be just one more adversarial proceeding, they failed to get any authority from the tribal council to make a commitment to any agreement that did not cancel the proposed plants. They also failed to bring along a Cheyenne who could adequately represent the tribe and offer new perspectives on the principles motivating the tribe's position. When the representatives of Montana Power and the EPA realized that these representatives had no power, the negotiations collapsed. Months later, discussions between Montana Power and new bargaining representatives of the tribe—Cheyennes endorsed by the tribal council to represent their views—succeeded in developing an agreement where other discussions had failed (Sullivan, 1984).

Failures of leadership can stem from a failure of organizational structures to support bargaining or endorse individuals as legitimate representatives of a collective interest. In the Iranian crisis, the power to speak on behalf of the Iranian nation (and the ability to influence the captors) rested outside the usual channels of government. Offices and titles that seemed to confer power did not. In the Montana Power–Northern Cheyenne dispute, no institution or structure designated individuals to bargain for the tribe in the EPA proceedings. The tribe sent the same legal counsel that had presented their position to other regulatory proceedings. The Northern Cheyennes' participants at the EPA-sponsored negotiations did not even believe that bargaining would take place.

In a dispute over a development project, no laws or institutions endorse individuals as bargaining representatives of the conflicting groups. Negotiations often lack a formal legal status. Existing laws enhance the power of individuals to litigate rather than to negotiate, and environmental laws often channel a dispute into a judicial forum. The action of any one individual can bring a lawsuit, while negotiation requires the assent of almost all participants. In addition, the outcomes of court suits carry legal authority. Negotiated agreements can have an undercertain legal status (Kretzmer, 1979). Without the endorsement and binding authority of law, negotiated agreements must rely on the goodwill of the negotiators for fulfilling the agreement terms. Since negotiations only follow a period of conflict and the bargaining settlement can compromise issues of major concern, it can prove particularly difficult to muster trust for an agreement that may have little legal force. Nevertheless, even without institutional endorsements, imaginative bargainers can develop arrangements that enhance their creditability. In the White Flint Mall dispute, the department store developer put aside a bond to create a self-enforcing agreement (Rivkin, 1977).

Institutional and legal structures also create issue agendas, determine the frequency of negotiations, and moderate the relationship between the disputants. These aspects of a relationship can either discourage or foster bargaining. When structures or traditions fail to support bargaining, then even the existence of desirable settlements and the best wishes of the bargainer can fail to produce an agreement.

Communications Failures

Communication skills can facilitate the cooperative search for a negotiated settlement. To bargain effectively, one must have not only the ability to articulate interests and bargaining positions, but also the skill to interpret and transmit bargaining communications to the other negotiators. The lack of these skills can discourage groups and individuals from bargaining or cause the bargainers to make errors that reduce or eliminate the ability of the negotiations to produce a settlement.

A first step in bargaining is to establish procedures for handling routine bargaining communication, arranging meeting times and places, choosing the bargaining agenda, and recording progress on different issues (Simkin, 1971; Cormick and Patton, 1977; Young, 1970). Negotiators may disagree over procedures for a variety of reasons, including aversion to bargaining, a belief that certain procedures advance their interests, a desire to use a disagreement over procedures as a surrogate test of will, or a strategic opportunity to delay (Iklé, 1964).

The peace talks between the United States and North Vietnam began with a struggle over the shape of the bargaining table that dragged on for 13 months. This disagreement probably continued for so long because of all four factors. Neither the United States nor Vietnam trusted each other and both were reluctant to bargain. The shape of the table carried symbolic meaning concerning the status of the Viet Cong and the South Vietnamese government. This initial test allowed both sides to show their willingness to bargain aggressively. Finally, concerns for battlefield events and domestic elections usually made one side reluctant to bargain when the other was willing (Kissinger, 1979).

Without some mechanism for routinely solving procedural issues in a fair and expedient manner, bargaining can stall as each side avoids advancing to a comprehensive review of interests or to the difficult give-and-take of meaningful negotiations. When groups have a history of bargaining together, past practices can establish patterns that bargainers use in future negotiations to avoid disagreements concerning meeting times, bargaining procedures, or bargaining agendas. First-time bargainers must develop their own procedures for resolving these routine bargaining matters. The lack of routine procedures can add to the anxieties that first-time bargainers have as they enter into bargaining. In development disputes, bargainers generally negotiate with each other on a one-time basis.

Fear and mistrust can inhibit the ability of disputing groups to reach a settlement even when there are ample opportunities for designing settlements that can leave each side better off than a continuation of adversarial struggle (Deutsch, 1973). When great asymmetries of power separate the groups locked in a dispute, then face-to-face negotiations may present special problems. If the fear and anxiety that stem from real asymmetries in power parallel asymmetries in bargaining skill, then anxiety over communications itself may preclude effective negotiation. This impediment to bargaining stems both from the psychology of interaction and from the realities of the bargaining relationship (Deutsch, 1973).

Any agreement requires an element of trust. Without some trust, it is difficult to create a climate that enables the frank discussion of interests or the exploration of potential agreements. Bargaining messages also reveal how a group values certain issues (Pruitt, 1981). The strategic value of this information can enable one side to improve its bargaining position. Unless the negotiators trust that bargainers will not retract accepted offers, they will make them only reluctantly if at all.

Similarly, those who overestimate their power may see little need to negotiate or to make concessions. Overconfidence can present as great an obstacle as fear and anxiety. When bargainers are able to recognize the existence of joint interests that link them together and see that joint action will provide a better outcome than individual action, then agreement is possible. In the dispute over constructing power plants at Colstrip, Montana, uncertainties created by the passage of the Clean Air Act enabled both sides to confidently believe that acting alone they could gain everything that they desired. Montana Power believed that the clean character of their plants would necessitate that EPA give them a permit, while the Cheyenne Indians believed that the ability to redesignate the air over their reservation would enable them to cancel any nearby project that they found unacceptable. Only after years of litigation did both parties come to understand that new laws prevented unilateral actions from succeeding and that they both had an interest in ending the dispute through an agreement.

Many agreements require future actions. To some degree, both sides in a dispute must trust that the other will honor these promises. This can prove especially true in international relations, where there seldom exists any higher authority that can compel a nation to honor the terms of its treaties (Iklé, 1964). Even in domestic disputes, where injured parties have access to courts, contractual agreements cannot envision all contingencies and justice is neither swift or certain. Groups will accept agreements that require future actions only when there exists a basic level of trust (Deutsch, 1973; Simkin, 1971). Without some trust between the parties, negotiations cannot hope to succeed.

Conclusion

Our brief discussion indicates that negotiations over environmental disputes can fail for a wide variety of reasons, including:

- Passions and principles can preclude a productive discussion of interests or potential agreements.
- Struggles can win new rights.
- A group may have little to gain through negotiations.
- The underlying interests of the competing groups preclude consensus solutions.
- Uncertainties foster an unrealistic assessment of a group's strength.
- Unstable groups cannot stand the stress of bargaining compromises or the discussion of competing interests.
- The wrong individuals negotiate.
- Laws promote litigation.
- Bargaining communications are difficult.
- Procedural pitfalls stall bargaining.
- Fear of bargaining precludes settlement.
- Insufficient trust exists to implement settlements or support bargaining promises.

This list suggests that negotiations can fail to produce a settlement for a wide variety of reasons, many of which stem not from differences of interests, but rather from the limitations of groups, leaders, and procedures, and from communication difficulties that characterize many disputes.

The next chapters will examine the more mature negotiation contexts of collective bargaining and international relations. These chapters attempt to determine which elements of group and institutional structures most enable negotiators to substitute constructive dispute resolution for more destructive adversarial contests. In addition, these chapters will attempt to determine how mediators affect the ability of groups to reach negotiated settlements. This analysis should help us to determine how one could design institutions and processes for resolving disputes surrounding development projects in less wasteful and litigious ways.

CHAPTER 3

Groups in Negotiations

Introduction

Groups interact in settings that shape both the group and the negotiations. Further, the internal structure of a group affects its relations with the external world and its ability to negotiate. This chapter examines the effects of bargaining environment, group structure, and group leadership on negotiation and mediation. An analysis of the highly structured setting and groups of industrial relations, the formal setting of international affairs, and the less structured settings and groups of environmental-development disputes shows how society, custom, and law can shape an organization and its ability to bargain effectively. The disputants in these settings include a diverse range of groups. They include labor unions, management associations, national governments, and advocacy groups that differ in their objectives, leadership structure, and membership rules.

Society has at times acted to alter the internal structure of groups to facilitate the negotiation of disputes. In particular, an examination of the development of industrial relations shows how government has acted not only to foster collective-bargaining in labor–management relations, but to regulate the internal structure of labor unions, both to facilitate the negotiation of disputes and to protect the rights of workers from abuse by union officials. The history of government action in industrial relations suggests how laws might enhance a group's power, authority, and ability to negotiate the resolution of a dispute.

Mediators have aided diverse disputants in all three of these settings to reach negotiated settlements. This chapter explains how the mediator's tasks vary with the settings; with the membership structure of a group; with the skills, powers, and interests of the negotiators; and with the range of differences within a bargaining group.

Varied Settings and Actors in Negotiations

Negotiation and mediation take place in a variety of contexts between quite different actors. The success of a negotiation may depend not only on the issues that separate the conflicting parties, but on the nature of the setting within which the bargainers must resolve them. Effective negotiation or mediation requires that one understand the character of the relations between groups, its history, and the constraints of law and custom that structure the bargaining.

Collective bargaining offers a formal system of compulsory bargaining. The key factor in the relationship between labor and management is its formal and continuous nature. Once a union earns certification by the National Labor Relations Board, law requires both management and labor to resolve their disputes through periodic negotiations. Furthermore, labor and management meet with each other daily in the workplace. The agenda of bargaining is set and constrained by law within broad limits. Over time, relations develop a history. The negotiating parties eventually know each other well, and all issues are embedded within the ongoing relations established in the workplace. Labor and management bargain not only about wages and benefits, but also to determine the rights and privileges that each has in the workplace.

The history of this interaction, both on the job and in past negotiations, simplifies the task of communicating bargaining demands. Parties generally dispute real issues, and only rarely perceptions. When the parties share a history of cooperation, they need little assistance in managing bargaining communications or negotiation procedures. Good communications and effective bargaining tend to dominate U.S. labor relations. Collective bargaining and mediation have enabled labor and management to generally avoid damaging strikes. Strikes account for less than 0.3% of the total person-days of nonfarm working time (Simkin, 1971). Strikes hurt both labor and management—labor loses wages and management loses production. The small number of days lost in strikes suggests that labor and management do avoid destructive conflicts through negotiation.

Laws have institutionalized the collective-bargaining relationship and determined its broad features. Although Americans take for granted the collective-bargaining process that characterizes organized industrial relations, the current system is a result both of the bargaining experience of management and unions and of the historical development of a system of laws that has modified industrial conflict and supported collective bargaining. Government policy and law now shape both the setting of labor–management relations and the internal structure of unions.

American public policy toward labor unions has varied from oposition, to support, and to control. Herman and Kuhn (1981) characterize 1800–1932 as a period of opposition, 1932–1947 as a period of support, and 1947 to the present as a period of control. From 1800 to 1932, the courts viewed strikes and other union activities as conspiracies. The Sherman Anti-Trust Act of 1890 states that "Every contract, combination in the form of trust or otherwise, or conspiracy, in restraint of trade or commerce ... is hereby declared to be illegal." From 1890 to 1932, this act affected not only big business, but also labor's attempts to act collectively. Courts issued injunctions, which proved a potent weapon against unions (McNaughton and Lazar, 1954).

The economic disaster of the Great Depression led to 15 years (1932–1947) in which federal legislation supported labor's attempts to organize and bargain collectively. In 1932, the Norris-Laguardia Act denied courts the power to forbid strikes and limited the grounds for issuing injunctions. The National Industrial Recovery Act of 1933 (NIRA) attempted to halt the downward spiral of wages by actively encouraging the creation of unions. This act stated that employees had the right to organize to bargain collectively and that employers could not limit an employee's right to join a union. This act, however, proved almost unenforceable and was ruled unconstitutional by the Supreme Court in 1935 (McNaughton and Lazar, 1954).

The National Labor Relations Act (Wagner Act), passed by Congress in 1935, duplicated the NIRA's endorsement of collective bargaining but developed an effective (and constitutional) framework for protecting workers' rights through the power of a National Labor Relations Board. This act not only enhanced the power of labor to organize but also began to control the methods of organizing and to influence the structure of unions. Under the provisions of the act, management had to recognize and bargain in good faith with any union that won a secret ballot election among the workers in a bargaining unit. The reliance on election procedures began to inject a democratic structure into union organizing activities. Labor historians view the Wagner Act as the high point of public policy endorsement of unions (Herman and Kuhn, 1981; McNaughton and Lazar, 1954; Ullman, 1955).

Shortly after, the mood of government and the public changed from support of organized labor to concern over union power and union abuses of workers. The Taft-Hartley Act of 1947 outlawed jurisdictional strikes and sought to check a union's abuse of workers under closed-shop agreements, including unreasonable expulsions of workers by the union or the refusal of union membership to workers (McNaughton and Lazar, 1954). This act also created the Federal Mediation and Conciliation Service to help parties to settle their disputes, and it gave the president the power in national emergencies to initiate a fact-finding process that could lead to an injunction against a strike.

In 1959, Congress passed the Labor-Management Reporting and Disclosure Act (Landrum-Griffin Act), hoping to curtail abuses of power and corrupt practices within unions. Although the extent of corruption that existed was likely small (Bok and Dunlop, 1970), the act took strong steps to insure the public disclosure of the terms of negotiated contracts, loans to union officials, and expenditures of funds to influence workers. The act also requires fair, honest, and frequent elections of union officials and thereby imposes a democratic structure on the internal organization of unions and attempts to insure the fairness of election procedures.

This series of laws shows how public policy has shaped the present structure of labor–management collective-bargaining and the organization of individual unions. The changes in legislation transformed the very character of the labor movement by greatly enhancing the ability of workers to form large industrial unions with democratic organizational structures. Between 1933 and 1939, union membership went from 2.9 million to 9 million (Ullman, 1955). Industrial unions rose in power and numbers. By 1940, the newly formed Congress of Industrial Organizations (CIO), which was organized by industry, had a membership of 3.6 million compared to 4.2 million in the long-existing American Federation of Labor (AFL). The rapid growth in the CIO became possible because organizing elections in large workplaces offered a simpler way of rapidly increasing membership than organizing by craft or skill, the technique then used by the AFL.

A mature collective-bargaining relationship, shaped by custom, history, and law, also simplifies the negotiators' and the mediators' tasks. The Taft-Hartley Act established the Federal Mediation and Conciliation Service and established a notification process that facilitates mediator intervention in stalled contract talks and funds mediation efforts (Simkin, 1971). Further, the ongoing relationship between labor and management that is part of mature collective bargaining reduces the communications obstacles that bargainers face. Only about 11% of labor–management relations require any mediation (Simkin, 1971). In these disputes, mediators find that bargaining often fails because of bargaining tactics and issues, rather than from major failures of communication.

International relations offers an example of a formal, noncompulsory system of bargaining in which no laws regulate conflict between nations, and where communications barriers arise from differences in culture, language, and government. Unlike collective bargaining in the United States, international negotiations take place only when the conflicting nations desire them. Thus, countries need not bargain to a negotiated settlement or establish ongoing relations that can routinize bargaining practices and reduce the potential for misunderstanding. War remains an alternative, and a look at a newspaper suggests that it remains a common one.

Even when countries have an ongoing relationship (and even the same language), cultural and national differences can easily create failures of communications that can complicate bargaining. In *Alliance Politics*, Richard Neustadt (1970) shows how "muddled perceptions, stifled communications, disappointed expectations, and paranoid reactions" (p. 56) produced diplomatic crises in British-American relations in 1956 and again in 1962. In 1956, the United States pressured Great Britain into accepting a ceasefire that prevented its capture of the Suez Canal (which Nasser had nationalized). In 1962, Kennedy canceled a $2.5 billion Skybolt weapon system on which Great Britain had placed its defense hopes. In Neustadt's view, both these incidents grew into crises as government officials assumed that their counterparts understood both that the nation's internal politics constrained action, and that their counterparts faced similar incentives and constraints from their domestic politics.

Despite the long history of U.S.-Great Britain cooperation that was forged in World War II and the North Atlantic Treaty Organization (NATO), these two crises quickly eroded good feeling. In the Suez crisis, Great Britain almost broke diplomatic relations with the United States over American actions to force them to abandon Suez. In the Skybolt affair, only the substitution of the most advanced weapon system, the Polaris nuclear submarine, permitted Kennedy to defuse Britain's anger over his public depreciation of the capabilities of the Skybolt weapon. Neustadt's book concludes with an acknowledgment of the difficulty and need of understanding the forces affecting the perceptions of government officials in different but allied countries.

When negotiations take place between nations without a past bargaining relationship, or in a climate of hostility, then communications problems prove a major complication of bargaining. In the Iranian hostage crisis of 1979–1981, many failures of communication and much distrust marred the earliest attempts to negotiate a hostage release. Only after months of negotiating with government officials did the United States realize that only Ayatollah Khomeini, the religious leader of Iran, possessed the power to release the hostages. This misperception arose in part because Americans seemed to expect that governmental titles in Iran carried the power to act.

When communications problems complicate bargaining, mediators who understand both cultures can help establish a constructive bargaining relation (Fisher, 1969). Unfortunately, the participation of a mediator in an international dispute is not set by any particular procedure. Often, mediation takes place only after a crisis is reached. International organizations, such as the United Nations, can urge conflicting countries to accept individuals or committees to mediate a dispute, but no sovereign nation need accept mediation (Young, 1970). In the Iranian hostage crisis, for example, two French lawyers and the Algerian government helped establish a bargaining relation (Salinger, 1981) while United Nations efforts were rebuffed.

Environmental negotiations over a development project offer a less structured setting for resolving disputes. Currently, the term *environmental negotiations* refers to a set of many different activities. In any of these negotiations, individuals meet representing environmental, antidevelopment, industrial, and government interests. These individuals and their groups seldom have a past bargaining relation or a history that establishes channels of communication. Both the developer and the regulatory authorities will have hierarchical organizational structures, but they can differ greatly in their ability to make decisions. Entrepreneurial developers can make decisions that are subject only to the discipline of the market and their financial backers. Bureaucratic government agencies will find their authority circumscribed by law and procedures. Although disputes take place within the same country and community, some have argued that developers come from a culture that sees risk as offering opportunity while opponents often see risk as posing harms to avoid (Susskind et al., 1978; Douglas and Wildavsky, 1982).

Groups opposing a facility may lack both the organizational structures needed for making decisions and the authority to act as a group. When a group forms to oppose a particular facility, the early history of a conflict generally includes bitter confrontations aimed at winning the recognition of the legitimacy of the concerns championed by those who oppose a facility. In many situations, only the common interest in opposing a particular project links the individuals to a group. There are few laws or customs that structure the relations of the individuals or groups with each other. The few existing laws encourage litigation, not negotiation.

Current attempts to resolve disputes over development projects occupy a legal netherworld. Attempts to negotiate the resolution of a dispute over a development project lack the endorsement, structure, and legitimacy that laws give collective bargaining. Although all the disputants in a development conflict reside within the same community, without some legal recognition of opponents as a group bargainers face extreme difficulty in designing settlements that will not spawn litigation that undoes the agreement. If every dissatisfied union member could initiate litigation on the terms of a work contract, a collective-bargaining agreement would have much less value. In development disputes, the legal structures that facilitate the intervention by any citizen in a regulatory proceeding discourage attempts to bargain with any group.

A mediator has generally assisted in those environmental negotiations in which the disputants have reached a negotiated settlement. In these situations, mediators have worked to establish routine communications and to overcome bargaining obstacles caused by the unfamiliarity of the groups with bargaining.

Table 1 summarizes our comparison of the settings that characterizes these three negotiation systems. U.S. labor–management offers a formal bargaining setting delimited by law in which labor and management meet in an ongoing relationship. Both parties share the same culture and language. The most bitter conflicts generally reside in the distant past. Government provides mediators when bargaining fails to produce agreement.

Current international relations offers a bargaining setting where protocol and tradition form the bargaining setting. Nations negotiate in bilateral or multilateral settings. Both the nature of the relationship between nations and the history of a dispute vary widely. Language and culture generally differ. The participation of a mediator in many disputes arises from ad hoc arrangements, but international institutions such as the United Nations will provide mediators in times of crises.

Development disputes lack formal bargaining structure. Disputes can involve developers, government agencies, and community groups, but both the role and the number of disputing groups vary greatly. Although parties share the same language, attitudes toward development suggest cultural differences. Rarely do disputants have an ongoing relationship, and the immediate history often includes aggressive interaction. Mediators are involved on an ad hoc basis, and no independent institution routinely sponsors them.

A brief review of Table 1 suggests that the setting of development disputes often has more in common with international relations than with domestic industrial relations. In the Snoqualmie Dam dispute (mentioned in Chapter 1), which is in many ways typical of disputes resolved through negotiation, bargaining took place at the special urging of the

Table 1
Settings and Actors in Industrial Relations, International Relations, and
Development Disputes

	U.S. industrial relations	International relations	Development disputes
Structure of interaction	Formal bargaining relation. Interaction set by law. Compulsory bargaining.	Formal bargaining relation. Interaction set by diplomatic protocol and tradition. Bargaining by choice.	No formal bargaining relation. Interaction set in adjudicatory forums. Bargaining by choice.
Actors	Unions and management in a bilateral relationship.	National governments. May be bilateral or multilateral.	Developers, regulators, and community and interest groups. Commonly multilateral.
Relationship of disputing parties	Ongoing.	Depends on the nations.	Often a one-time relationship.
Communications and culture	Same.	Often differing languages; always a different culture.	Same language, but differing subcultures.
History of conflict	The most bitter disputes are often part of distant past.	Varies with the negotiations.	Conflict history often includes recent bitter interaction.
Mediator	Formally involved by law.	Ad hoc, but international institutions such as the UN often provide assistance.	Involvement often ad hoc.

governor and involved environmental groups, local farmers, and government. The parties lacked any ongoing relations, and came from very different backgrounds. The recent history had been marked by litigation and catastrophic flooding, which served to accentuate differences. The mediator entered the dispute at the governor's urging, and he helped structure the relations between the parties. The lack of bargaining structure and ongoing relationship between groups complicated a discussion of underlying interests and the search for settlements that offered mutual gains.

Laws that endorse and structure bargaining between groups proposing and opposing a particular development project could greatly enhance the likelihood that environmental negotiation would resolve the conflicts. Just as the development of labor law marked a conscious policy choice to encourage and regulate collective bargaining, laws could endorse and support efforts to resolve environmental disputes through negotiation. Just as laws endorsed union democracy, law could create incentives that enhance the bargaining ability of groups intervening in a development dispute and could strengthen a bargaining framework that would permit wider searches for settlement.

The tasks of a mediator also vary greatly with the negotiation setting. When parties lack a relation with a past or future, then mediators must work to overcome the communications obstacles as they arise. In bargaining situations with mature and ongoing relations, such as collective bargaining, negotiators need mediator assistance only rarely. Here, a shared history affects both the definition of issues and the levels of animosity or trust that characterize bargaining. Although a history of acrimonious relations may increase the emotional level of interaction, past relations generally reduce communications problems.

When no law or custom works to control bargaining interactions, a mediator may need to establish a framework for negotiations as a precondition for bargaining. Thus, an understanding of the relations between groups, their history, and the constraints of law and custom that moderate interactions is an important starting point for those seeking to mediate disputes. Environmental and international mediators often find that they must first discuss the ground rules for negotiation before discussing issues of substance. Subsequently, mediators attempt to furnish the bargaining setting with structures that law, custom, or prior history fail to generate.

Membership Structures and a Group's Ability to Bargain

The membership structure of the group's whose assent is essential to the implementation of an agreement can critically affect the outcome of negotiations. This membership structure can range from legally required association in a single group to voluntary association in many groups.

To bargain effectively, an organization must maintain a stable leadership coalition capable of articulating the group's interest and formulating bargaining proposals. A successful group must maintain its membership and resources not only during bargaining, but over time. The articulation of group interests and generation of bargaining proposals can threaten both the stability of the leadership coalition conducting bargaining for the group and the long-term stability of the group itself. When membership in a group is compulsory, then the threats to the organization from bargaining are small. Bargaining leaders, however, must worry about their own ability to maintain the support of their leadership.

In voluntary associations, bargaining can pose a threat not only to the current leadership coalition, but also to the long-term survival of an organization. Wilson (1973) argues that four forces hold organizations together: "material incentives" such as the pay and profits that hold together a large business; "specific solidarity incentives" such as the ranks and titles that supplement the low pay in the military and in academia; the "solidarity incentives" that arise in part from group interaction and self-esteem; and "purposive incentives" that arise from the satisfaction that members get in working toward a common goal. In voluntary associations, which cannot rely heavily on either material incentives or titles and honors, the esteem of fellow members and the commitment to a common goal play large roles in maintaining a group.

The articulation of interests and the formulation of proposals for bargaining can adversely affect both the solidarity and the purposive incentives that bind members together. To arrive at bargaining settlements that offer both sides gains, negotiators often must articulate underlying interests and explore alternatives that differ from positions previously espoused. Such an articulation of interests may be viewed by some members of the group as a departure from the original goals and positions. This perception can produce an assessment by members that may threaten either the position of the bargaining leaders who articulate these interests or, if members drop out, of the group itself. Even though voluntary associations may aim for a life that lasts only as long as the dispute, the formation of splinter groups can undermine the bargaining effectiveness of the original group and the value of negotiations itself. When bargainers must not only articulate interests but must compromise a group's positions, then the threat to the group's leadership and stability is even greater.

Labor–management negotiations take place between groups whose members cannot readily abandon them. A certified union represents all workers in a shop whether or not they belong to a union (Olson, 1973). In addition, 95% of all unionized labor has contracts that include a union security clause which requires workers who benefit from union contracts to pay dues, even if they choose not to join the union (Olson, 1973). Thus, by law, a union represents all workers in a shop, and in practice, almost all workers in a unionized shop belong to the union.

Management membership in one bargaining team is determined by business policy. The production department, for example, cannot secede from a firm and reach a settlement while the marketing department continues to bargain. Dissident managers must resolve their differences within the firm's management structure or else leave the firm entirely.

Law stabilizes the membership structure in labor–management relations in ways that enhance the ability of labor and management to bargain. Law creates a bilateral bargaining relation between a labor union that represents all workers in a shop and management. The laws that determine membership in a union simplify both the negotiation task and the mediator's role. The bilateral structure eliminates the need for bargainers or mediators to worry about unstable coalitions, multigroup dynamics, or splinter groups that complicate the task of achieving an agreement (Raiffa, 1982). Once a negotiated settlement receives ratification by a vote of union members, all must honor it. Laws help insure that dissidents will not undermine negotiated and ratified contracts. Laws prevent labor dissidents from creating a competing union to simultaneously bargain with management (for only one union can represent a worker). Law enables both labor and management to seek judicial enforcement of contract terms. Thus, negotiators and mediators may generally presume that a ratified settlement will be an honored agreement.[1]

In international relations, nations generally accept governments to speak for the citizens living within a territory.[2] All citizens in a country are bound by the actions of a government in its conduct of foreign policy. Those who disagree face the power of the state. Dissenting individuals have few choices other than to work for a change of national policy or government, to refuse to comply with government actions, or to eliminate the existing government.

Although mediators in international conflicts must remain sensitive to the domestic politics of each country, bargaining stresses can rarely lead to a government's collapse. The negotiations to resolve the crisis resulting from the seizure of U.S. diplomats by Iranians is a noteworthy exception to this general rule. The negotiating secular leaders had little real authority, and the act of bargaining with the United States left them open to criticism from religious leaders that could undermine their government itself. Salinger's *America Held Hostage* (1981) documents how Banisadr, President of Iran, and Ghotbzadeh, Foreign Minister, had to repeatedly back away from bargaining concessions and offers because of their fear of denunciation by the powerful Iranian clergy. Rarely do leaders of government possess so little power.

---

[1]Wildcat strikes are the exception to this rule.
[2]Some nations will at times refuse to recognize governments as the legitimate representatives of a people.

In environmental-development disputes, groups opposing projects often are voluntary associations that possess quite loose organizational structures. Membership is not limited to one group that must embrace the interests of many. Although membership is always voluntary, the rules for membership can vary widely. Some groups may require only agreement with their views, while others may require dues payment and sponsorship by other members. The Clamshell Alliance, which opposes nuclear power plants, keeps no formal membership lists or leadership structure. The Sierra Club, on the other hand, uses elaborate mailing lists to generate support and attract members to the group.

Unlike unions, no single group is certified to exclusively represent individuals, and individuals may join several groups. Many others will belong to none. Each group may either represent a single interest of major concern to its membership or embrace several interests. These voluntary and informal structures make it difficult for a leader to commit a group to a specific course of action. The leadership often does not represent its members but simply promotes consensus positions on issues. When membership springs from shared positions, then a reformulation of basic interests or compromise can threaten the stability of the leadership, the membership, and the group's resources. Since those leaving a bargaining group are free to initiate litigation, they possess great power.[3] This ongoing threat of litigation reduces the value to developers of reaching an agreement with the remaining bargaining participants.

On the developer side of these disputes, membership questions are more easily settled. Just as in industrial relations, private sector firms can clearly resolve who belongs in their group for negotiations. A single company usually possesses responsibility for all major aspects of production and can effectively resolve all internal disagreements. However, when government is the developer, then at times a decision to include features in a project to mitigate impacts on local communities may require the assent of several levels of government or different government agencies. The controversial project to rebuild the West Side Highway in New York City, for example, requires the agreement of the mayor, the governor, the U.S. Department of Transportation, and the U.S. Environmental Protection Agency to a particular project design. For many years, political struggles have produced no decision either to abandon the project or to start construction. Although membership in any government group is clearly determined, bargaining becomes multilat-

---

[3]See the Englewood Metal Recycling Case in *Environmental Mediation: An Effective Alternative?* (1978) for an example of how the refusal of some to join a negotiation effort caused its failure.

eral, rather than the simpler bilateral relation common in labor–management relations, and new issues constantly arise.

The membership structure of competing groups greatly affects the mediator's role. The compulsory membership that dominates labor–management and international relations simplifies the task of the mediator. Bargaining can proceed without fear that negotiation stresses will generate a new group that can subvert the bargaining process. Nevertheless, even in groups with compulsory membership, a mediator may help a bargaining group to resolve internal issues (Kerr, 1954; Stevens, 1963). This task is much simpler than obtaining a consensus of the group's membership on every issue and bargaining position.

In resolving disputes involving groups with membership characterized by voluntary association, a mediator must pay particular attention to the internal dynamics of the bargaining groups. When a group includes only members with a single interest, then negotiation can pose a risk that a new group may split off to champion that issue if bargainers consider compromise. The mediator, in these situations, may need to help a group maintain an internal consensus. When a mediator is well known and has the respect of the membership of competing groups, then his or her endorsement of a pact may help insure the acceptance of a negotiated settlement.

Skills, Interests, and Powers of Bargainers

To facilitate orderly bargaining, groups generally designate individuals or teams to represent their interest in negotiations. The effectiveness of bargaining in resolving disputes increases when the negotiators possess the bargaining skills that enable them to effectively articulate their interests, a direct personal interest in bargaining, and the authority to make bargaining demands, concessions, and commitments. When negotiating skills, personal interests, or bargaining authority are lacking, it is almost impossible to settle a dispute. At times, however, a mediator may take action to enhance the ability of the bargainers to reach a settlement.

The range of bargaining skills, interests, and power can vary widely both within and across bargaining settings and organizations. In general, the laws and procedures that support labor–management negotiation have fostered the selection of skillful bargainers who possess both a strong personal interest in negotiating and a substantial bargaining authority. In international relations, diplomats are trained negotiators whose reputation depends on their professional bargaining skills. The ongoing character of both industrial and international relations enables the bargainers to develop both negotiating skills and an understanding of the bargaining process that is often lacking in environmental dispute resolution. When a setting does not develop the skills and powers of bargaining representatives, a mediator's actions can sometimes help groups overcome some of the obstacles that these deficiencies present to the search for a negotiated settlement (Stevens, 1963; Pruitt, 1971; Douglas, 1962).

In labor–management relations, negotiating skills bring benefits to the group and the individual bargainer. In turn, these skills support the negotiation process. Bargainers win reputations within their organizations based on their bargaining skills and the settlements they win. These skills are shared throughout a union, and large labor unions will provide their locals with skilled negotiators. Similarly, large businesses will have industrial relations specialists who are noted for their bargaining skills (Slichter et al., 1975). The skills of these bargainers can help the negotiations to avoid bargaining pitfalls that arise through failures of communication or inability to understand bargaining signals. Thus, the bargaining setting rewards negotiation skills, and the skills support bargaining.

Bargainers representing labor and management generally possess the authority to make an agreement. In general, the negotiator's bargaining power will come from the power of the group he or she represents, and from his or her own control over the organization. To successfully negotiate, a group must have the power to affect the outcome of the conflict situation (Simkin, 1971). A bargainer must also possess the ability to represent and to speak for the members of his or her group. The participation of high company and union officials in contract talks has become common practice (Simkin, 1971). Each side can readily assess the power of the other negotiators through relatively clear signs, such as the position of a negotiator in the management or union hierarchy, the length of the official's tenure, and previous experience in dealing with the negotiator.

The authority of a negotiator can dramatically affect the bargaining. When a negotiator lacks authority to make a deal or compromise, negotiations can languish as the futility of the endeavor becomes apparent to all. Sometimes, however, a negotiator who lacks authority may attempt to win extreme gains to enhance his or her position within the organization, or to help advance his or her career. Thus, when negotiators lack the power and authority needed for bargaining, extreme demands can replace reasoned bargaining. In either case, the prospects for a negotiated settlement diminish with a bargainer's lack of power.

Power, and its lack, has strategic implications for any settlement. Schelling, in *The Strategy of Conflict* (1960), discusses the effects of power for deciding the outcome of a conflict. He states that the lack of power to make a concession can sometimes prove a powerful weapon in conflict. Nevertheless, unless negotiators possess a minimum amount of authority and power that enables them to speak for the people whom they represent, negotiations will prove futile. Although the procedures for ratifying a contract vary by union, effective union leaders can commit their membership to the contract terms. If management does not believe that union negotiators can either sell a contract to their members or lead an effective strike, they may see no reason to bargain with the union negotiators but may see the negotiations as a means for creating proposals for a membership vote (Summers, 1967). In 1978, the leadership of the United Mine Workers national negotiating committee lacked the ability to win the endorsement of the majority of its members. This situation caused the collapse of industrywide bargaining and led to a series of negotiations with the leaders of local unions.

The legally mandated membership and representative structures of labor unions usually hold a union's leadership accountable to all unionized workers in a bargaining unit. A union leader must represent and balance the diverse interests of union members in dealing with management. At many times, labor leaders must negotiate a compromise within their own organization between internal competing interests, such as the wage demands of young workers and the pension demands of older workers. If leaders fail in their balancing efforts, membership can replace them in an election via a secret ballot. Thus, leadership power stems from the ability of a leader to form a majority coalition to support his or her positions and the contracts he or she negotiates.

In international negotiations, countries choose diplomats with the skills to represent their interests at the bargaining table. Under almost all circumstances, negotiation with designated diplomats presents the only way of directly bargaining. In addition to possessing bargaining skills, diplomats often receive both personal and professional recognition for their negotiating successes. The incentives to reach a settlement are so great that Iklé, in *How Nations Negotiate* (1964), charges that at times U.S. negotiators take such a personal interest in reaching a settlement that they fear that the failure of negotiations will taint them with personal failure. This fear can lead bargainers to actually mediate between their country and other countries, rather than to aggressively present the position of their country, which, of course, can weaken the terms of the settlement that the diplomats negotiate.

Although professional diplomats generally possess bargaining skills and face strong personal incentives to produce agreements, their authority to bargain can vary widely. This authority can vary both because of the bargaining authority given to a diplomat, and because of the structure of domestic government. Negotiators can range in bargaining authority from simple messengers who repeat approved negotiation positions, to heads of state who possess broad powers to commit a country to a course of action (Iklé, 1964). Governments and institutions also limit the power of a negotiator. Governments can range from those dominated by checks and balances that limit a negotiator's authority, to those dominated by the will of an individual or group that can rapidly commit the nation (Kissinger, 1977).

Most commonly, a diplomat enters negotiations with the power to make some concessions and the ability to influence his or her government (Iklé, 1964). In these situations, one can view the acceptance of terms by the negotiator as a tentative commitment of his or her government. When bargainers have this much power, then negotiations can produce agreements for formal ratification. More rarely, diplomats may possess no real bargaining power. Iklé points out that under Stalin, Soviet diplomats acted as simple messengers. In these situations, both Eastern and Western diplomats quickly lose interest in bargaining, and face-to-face bargaining serves little purpose.

Although one might expect that summitry, in which heads of state meet for direct talks, would prove most productive since the bargainers have the power to resolve differences and commit their nations, this is not true. Iklé (1964) states that this advantage is offset by several disadvantages. National leaders can seldom have the command of detail necessary to construct a specific agreement. In addition, the failure of summitry can carry the risk of national disaster and seriously jeopardize future relations. Thus, in practice, the details of final communiqués are usually resolved even before the start of a summit meeting (Kissinger, 1979).

The domestic political structure limits the power of bargainers in international negotiations. In democracies, domestic scrutiny of the negotiation process can limit the effectiveness of bargainers. Iklé (1964) notes that negotiators like to portray their bargaining movements as major concessions. This practice, however, can lead to domestic charges that the negotiators are conceding too much. Iklé points out that in test ban conferences between 1958 and 1961, the United States revised its demand for the verification of compliance with underground test limitations several times to bring the U.S. position closer to that of the Soviet Union. These changes, however, were attacked as unreciprocated and dangerous concessions by members of Congress. The U.S. negotiators subsequently denied that they were concessions and portrayed these changes in bargaining demands to domestic audiences as based on a better understanding of the science of nuclear testing. This claim reduced the worth of these concessions in subsequent bargaining with the Soviet Union, which repeated the State Department's domestic argument that the concessions were minor (Iklé, 1964).

Although bargaining representatives of authoritarian and totalitarian states often have to satisfy domestic constituencies, the country's leadership generally has greater flexibility to both formulate and modify bargaining positions since they seldom need to publicly explain their actions (Kissinger, 1979). In bargaining to conclude the Vietnam war, the United States and North Vietnam debated for three months on the shape of the negotiating table. The U.S. government faced severe domestic criticism for not readily making concessions on what was generally viewed as a minor point.

The level of bureaucratic development of a country can seriously affect the power of diplomats and the outcome of negotiators. Kissinger (1977) argues that extensive bureaucratization can limit a country from exploring new bargaining initiatives and suggests that the internal requirements of a bureaucracy can dominate policy positions. Furthermore, disparities of bureaucratic structure can complicate ideological differences between the industrial (and bureaucratic) nations and new and developing countries. Bureaucratic inaction appears as arrogant indifference.

In conflicts over the siting and construction of a development project, the leaders of the groups opposing a project commonly lack the skills, interest, and authority to bargain effectively. Many of the groups that oppose a development spring up to stop a particular facility. Leadership adopts strong, principled advocacy positions that attract members to a cause. These groups can face a severe problem in winning a hearing for their views. Overcoming political malaise and bureaucratic indifference necessitates the formulation of a strident and uncompromising position. Little in this history enables a group to develop the skills for bargaining or an interest in this technique. Furthermore, the advocacy positions can sharply curtail a group's power to make bargaining concessions.

The lack of bargaining skills among groups opposing a project can pose a difficult negotiation problem. In the dispute over the construction of a power plant at Colstrip, Montana, only after several years of litigation did the Cheyenne Indians opposing the power plant win recognition of the inevitability of their participation in siting decisions. Those who won the recognition battle proved unable to bargain. The Cheyenne Indians and Montana Power reached a settlement only after the tribal council replaced the advocacy staff with a special negotiating team including tribal council members with negotiating experience. Similarly, personnel within Montana Power who favored a negotiated settlement gained power and replaced those favoring a litigation and adversarial approach. Changing leadership helped both sides gain the ability to negotiate (Sullivan, 1984). In groups that lack the membership stability of an Indian tribe or a power company, changing leadership to facilitate negotiations can prove impossible.

In addition to the lack of bargaining skills, the leaders of groups opposing development projects often lack sufficient authority to bargain effectively with the proponents of large projects. The membership of a single-issue opposition group often grows because the leaders attract the support of followers to their positions. A leader's strength becomes tied to the promotion of certain positions, and this leaves little room for an exploration of larger interests, and even less room for compromise. Departures from previously exposed positions can lead to the abandonment of leaders by their constituents, and to the creation of newer, more orthodox groups (Hirschman, 1970; Douglas and Wildavsky, 1982). Thus, compromise threatens an organization that relies on purity of motives and shared beliefs to forge membership bonds. Although union leaders must face election competition from other members, the threat of abandonment of the union by workers is small. The abandonment of collective bargaining is almost impossible. Unlike union leadership, the power of an environmental negotiator to speak for his or her constituents faces limitations quite different from those in collective bargaining. The Colstrip Power Plant negotiations were aided by the fact that the Northern Cheyenne tribal council had legitimate authority to speak on tribal concerns. This feature, however, is quite rare in development disputes.

Once again, labor law provides incentives that encourage unions to elect leaders that represent their interests and possess the power and authority to act on their behalf. Environmental negotiations now occur under a legal framework that encourages litigation rather than negotiation. Legislators, however, could alter this legal framework as they have altered collective-bargaining, both to stabilize the membership and leadership of community groups concerned with the consequences of development and to create incentives that encourage negotiation.

Mediators can play a particularly important role in those disputes where negotiators lack bargaining skills or the power to produce a settlement (Kerr, 1954). In labor–management disputes, mediators have assisted first-time bargainers to develop constructive bargaining patterns

(Pruitt, 1971). Communications and procedural issues can subvert first-time bargaining, and mediators can prove particularly helpful in resolving these disputes. Similarly, in environmental-development negotiations, mediators have met with the disputants prior to bargaining to help each side become familiar with negotiation procedures. In international relations, a respected mediator can help diplomats avoid bargaining pitfalls that arise through communication failures.

In disputes where negotiators lack bargaining power, there is less that a mediator can do. When negotiators lack power because of constraints imposed by opinions within the group they represent, then mediation by a prominent individual or organization may provide an endorsement of the negotiation effort that wins a bargainer needed support. In international affairs, mediation by the United Nations or by the leader of a prominent nation can serve this purpose (Young, 1967).

To overcome deficiencies in bargaining skills, a mediator can educate a bargaining party prior to the negotiating sessions concerning the likely evolution of the discussions (Stevens, 1963; Pruitt, 1971). During negotiation sessions, the mediator may manage the meetings and bargaining routines to avoid disputes over procedural matters. Although experience is the best teacher of these skills, disputes over development projects will likely bring together some groups that possess little negotiating experience. A later chapter will discuss the specific actions that a mediator may take to facilitate the acquisition of bargaining skills and etiquette.

The Range of a Group's Interests and Bargaining Issues

In addition to the differences that separate negotiating groups, a group may contain many members and subgroups that possess varied interests and concerns. These concerns may enrich negotiations by providing a large number of issues for a bargaining agenda but may complicate the bargaining tasks of determining what concessions to make and what terms an acceptable agreement must possess.

Zartman (1977), Iklé (1964), and Druckman and Mahoney (1977) each state that when bargaining over many issues, initial bargaining focuses on developing an agreement on principles and then proceeds to a negotiation of details. Pruitt (1981) suggests that these stages serve two purposes: "organizing the intellectual effort of the bargainers... and dealing efficiently with basic differences in outlook" (p. 14). He proposed the hypothesis that the increasing complexity and interrelatedness of issues increase the likelihood that a discussion of principles will precede a negotiation of details. This hypothesis suggests that the complexity of issues by itself will generate a discussion of principles and interests. Since this is often the first step for transforming a negotiation from a competitive situation to one that offers both sides gains (Fisher and Ury, 1981), richer agendas are likely to facilitate the negotiation process.

In arriving at a negotiated settlement, not only must conflicting groups reach a settlement across the bargaining table, but each negotiating team or group must reach a settlement on the issues that separate the different subgroups that make up its membership. Just like the bargaining across the table, the bargaining within a negotiating team will likely involve some combination of positions that leave all subgroups better off and some that strike a balance between the interests of subgroups. In labor unions, it is the task of union leadership to develop a balance between the differing subgroups, and union contracts often reflect the electoral power of a constituent subgroup. When dissent within a group is large, the mediator can assist by convincing the negotiator's constitutents that the negotiator is defending their interests (Kerr, 1954), by publicly accepting responsibility for the proposal (Simkin, 1971), or by attacking the negotiator for being too tough (Shapiro, 1970).

In international relations, the team negotiating on behalf of a country may include representatives of subgroups that have different perceptions of the importance of particular issues to a country's interests. In arms limitations negotiation, for example, U.S. State Department personnel generally place more weight on a treaty's effect on international relations than do Defense Department personnel, who stress the importance of military issues. Thus, even within a country, bargainers must balance these differing views.

In an environmental-development dispute, those opposing a construction project may include a variety of groups with quite different positions. Although all groups opposing a project may bargain under a single umbrella, antidevelopment groups can vary greatly in their geographic base and area of concern. They can range from national groups with many interests to local groups concerned with the impacts of a specific facility. In a dispute between the New England Power Company and local groups over the siting and construction of a nuclear power plant in Charlestown, Rhode Island, local groups were primarily worried about the construction impact of the plant, but they also shared the concern of many regional groups over the use of nuclear technology (Sullivan, 1980). This pattern emerges in those siting disputes that include many groups. In those situations where many groups in which a single-issue dominates form a loose coalition to oppose a facility, any alteration in opposition may threaten unity. Unless a law creates groups that represent more than one issue and interest, then the ensuing single-issue bargaining creates a zero-sum game where one group loses what the other wins. In this particular case, the opposing groups won a total victory, and the New England Power Company abandoned its plans after spending $30 million on engineering fees, legal fees, and regulatory fees. In the Colstrip case, however, the tribal government embraced many groups that had differing interests, thus making possible a bargaining attempt that could explore whether some alternatives to continued litigation could fulfill the interests of both members of the tribe and Montana Power. When a loose coalition unites a variety of opposition groups, a mediator will face a difficult task in maintaining the coalition under the normal stresses of bargaining. A mediator must help the chief negotiator to maintain the group coalition.

Mediators face a complex task in resolving the disputes within a group that arise in these three settings. Encouraging a discussion of principles and interests both across the bargaining table and within each group may produce an environment in which mutual gains may make all winners (Fisher and Ury, 1981). When no easy settlement is readily discovered between the bargaining groups, then compromises will require internal discussion. Much of the mediator's effort will focus on achieving a new understanding within a bargaining team that will support new alternatives. When the bargaining demands of a group constitute a wish list that includes the pet issues of almost every subgroup rather than a comprehensive statement of interests, then bargaining will generate observable gains and losses for each subgroup. This action can preclude a search for a settlement that permits mutual gains. As the parties move closer to an agreement, a bargainer's concessions must sacrifice the interests of a particular subgroup. These bargaining moves are inherently divisive and make bargaining extremely difficult.

Conclusion

The structures of disputing groups and the conflict setting interact in a variety of ways in the diverse negotiation relationships of collective bargaining, international relations, and environmental negotiations and mediation. The development of labor law has produced a legal environment that supports collective bargaining. Laws not only legitimate collective-bargaining efforts but require democratic organization structures in unions. These representative structures create incentives that enhance a leader's ability to bargain and increase his or her accountability to workers. In international relations, history and diplomatic custom can interact to support bargaining. Currently, laws limit the potential of negotiations for resolving disputes over development projects. Legal and regulatory structures encourage adversarial confrontations between project proponents and opponents. Such types of interactions can preclude future attempts to negotiate settlements.

Table 2 summarizes the comparison of the organization and leadership structures of the groups commonly found in these differing dispute settings. A review of Table 2 suggests that the negotiation of develop-

Table 2
Group Organizations and Leadership

	U.S. industrial relations	International relations	Development disputes
Stability of membership	Unions—compulsory membership. Management—set by position in firm.	Generally stable.	Opposition groups—membership commonly voluntary and tied to espoused position. Developer—membership determined by job.
Leadership structure	Unions—democratic/ hierarchical. Management— heirarchical.	Varies greatly. Diplomatic corps generally hierarchical.	Opposition group—varies, but often charismatic leaders attract followers. Developers—hierarchical leadership.
Incentives acting on leaders	Unions—bargaining successes are rewarded. Management— bargaining successes are rewarded.	Diplomats often face strong incentives to negotiate treaties. Incentives for national leaders vary.	Opposition group—bargaining poses threats to leadership and group stability. Developers—face incentives to avoid delay.
Range of interests	Unions—must embrace all members and all interests. Management— comprehensive economic interests.	Diplomats generally must reflect the multiple interests of a nation or of the ruling coalition.	Opposition group—may consist of a single-issue constituency. Developers—must embrace general development interests.

ment disputes faces several structural obstacles. Groups opposing development projects seldom possess a membership structure as stable as those found in unions. Leaders at times do not represent membership but attract followers to positions that they promote, which can make the articulation of basic interests and the modification of positions difficult. Finally, groups opposing development projects need not embrace more than a single interest. This narrowness can reduce bargaining to a divisive and bitter exercise in which one side wins what the other loses.

The analysis of the organizational structures of groups in these three dispute settings suggests the following conclusions:

- Stable membership structures facilitate representative collective bargaining. Laws can create or support organizational structures that help groups to bargain. In U.S. labor–management relations, laws create stable memberships and require the election of leaders through a secret ballot. These requirements enhance both a negotiator's authority to bargain and his or her accountability.
- Long-term bargaining relations create rewards for those who develop bargaining skills and reputations for following the customs of the bargaining relation. In industrial relations and international relations, negotiators develop reputations based on their ability to achieve negotiated settlements. The existence of a corps of skilled bargainers also enhances a group's ability to cast a dispute into a framework that permits negotiation.
- Successful bargaining requires that negotiators have authority and power. In labor–management relations, laws enhance the authority of a bargaining representative and establish clear procedures that bind groups to negotiated settlements. In international relations, government titles and formal decision processes commit governments to negotiated settlements. Leaders of environmental groups rarely have either the authority or the power to commit their groups.
- When laws or institutions create broad comprehensive groups, a dispute will possess many facets. A rich bargaining agenda can facilitate reciprocal concessions but require that groups develop a process for achieving an internal balancing of interests.
- Environmental legislation creates few incentives for resolving disputes through negotiation. Rather, these laws create incentives for individuals or groups to marshall their resources to advocate positions in litigation.
- Environmental and planning laws that protect an individual's rights to challenge development decisions limit the ability of

groups to bargain for a collective or cooperative settlement. Unlike in industrial relations, bargainers representing groups opposing development plans have little power to bind their members to a negotiated settlement. Members often find no procedures for recognizing or protecting their interests within the group but must form separate groups to advance their interests.

- Environmental law, by offering broad opportunities for standing, creates powers for those advancing a particular single interest. As groups form to protect a particular interest, the narrowness of the issue agenda complicates bargaining.

Mediation, like any activity that seeks to help people to resolve differences without resort to force, can work best when mediators are sensitive to the setting that shapes group interactions and to the characteristics of the competing individuals and groups. An identification of the factors affecting a negotiating setting and the bargainers should help mediators to determine the character of their tasks as they enter a dispute. Our analysis of industrial relations, international relations, and environmental negotiations suggests the following:

- The history of a bargaining relationship affects both the need for mediation and the mediator's tasks. In ongoing relations with a history of amicable dispute resolution, the negotiating parties are less likely to need mediator assistance. A mediator, however, may facilitate communications between groups bargaining for the first time. When groups share a history of hatred or distrust, then a mediator may help the parties to resolve a particular dispute by controlling and moderating bargaining exchanges.
- When neither law nor custom promotes bargaining or controls negotiation sessions, then a mediator can help the parties to agree on procedures to moderate bargaining exchanges.
- The instability of a group's membership can complicate the task of reaching a negotiated settlement. When neither law nor charter provide organizational stability, the participation of a mediator may help hold groups together under bargaining stress.
- When the leaders or representatives of the bargaining groups lack negotiating skills, a mediator may help orient them to the routine of negotiations.
- When negotiators lack power to bargain, then the participation of a well-known mediator in negotiations may help negotiators to win from their group the power to make bargaining commitments.

CHAPTER 4

Factors in Bargaining That Affect Negotiation and Mediation Success

Introduction

The bargaining procedures and protocols that structure negotiations can differ as dramatically as the organizational structures of the groups involved in negotiations. These features of the bargaining environment can range in formality from those set by law or custom to those adopted on an ad hoc basis. In each bargaining setting, the presence or absence of certain features can affect the likelihood of bargaining progress. These key features include:

- The existence of both common and competitive interests linking the groups;
- How groups are recognized as participants in a dispute;
- The balance of power between the competing groups;
- The frequency of negotiations;
- The number of bargaining groups:
- The existence of bargaining deadlines;
- The existence of binding agreements to conclude formal negotiations.

When these factors are present in a particular dispute setting, they can facilitate the resolution of the conflict through negotiation. Law and custom can interact to inject these features into a particular dispute. In less structured bargaining settings, prenegotiation conferences can also enable groups or nations to resolve many issues. In other situations in which laws provide few features that support negotiations, a mediator can take steps to incorporate structures into the bargaining setting that facilitate the achievement of a negotiated settlement.

Cooperative and Competitive Interests in Disputes

Any relationship between groups will include a combination of common interests (in which both sides desire the same outcome for the same reasons), complementary interests (in which groups desire the same outcome for different reasons), and competing interests (in which both groups desire a different outcome) (Iklé, 1964). Together, common and complementary interests introduce cooperative elements into bargaining. Both academic and popular texts on bargaining and conflict resolution compete in telling stories to illustrate the interconnections of competitive and cooperative interests in familiar and humorous settings. Two of the better ones are:

> A husband and wife must decide on how to spend a Saturday evening. He wants to go to a baseball game, and she wants to go to the ballet. Both, however, want to spend the evening together. The husband would prefer going to the ballet with his wife over going to the baseball game alone. The wife would rather go to the baseball game with her husband than go to the ballet alone. (Based on Schelling, 1960; Fisher and Ury, 1981)
>
> Two children are fighting over a single baked potato. To end the argument, the father cuts the potato in two and gives each half. Only then does he realize that one child wants the potato skin, and the other the inside. (Based on Cohen, 1980)

These two stories illustrate the often overlooked fact that in many disputes a combination of competitive and cooperative interests link the opposing sides. In the Saturday evening dispute, the husband and wife may compete on where to go, but they must cooperate to insure that they will stay together. This situation has no easy solution. More optimistic writers generally solve this conflict by having the husband and wife adopt an alternate weekend strategy: the husband chooses on one weekend, the wife on the next. The most pessimistic writers have them stay home and watch television.

The mixture of competitive and cooperative interests may vary in any dispute. The charm of the potato story stems from a sharp juxtaposition. In general, dividing a good (or making a sale) creates a bargaining situation that highlights the competitive aspects of bargaining. The more one gets, the less the other gets. No conflict, however, really exists between the children, and the father's fair split solves a dispute that does not exist.

Individuals and groups negotiate most readily when they believe that a settlement will make them all better off than continuing to dispute. Each side has an incentive to cooperate with another in a search for a settlement that can leave them better off than will the failure to settle. Many of the popular books written on negotiation offer advice on how to accentuate the cooperative aspects of any particular dispute. Advice includes "inventing options for mutual gain" (Fisher and Ury, 1981) and developing a "win-win technique" (Cohen, 1980). These books suggest many creative techniques for introducing cooperative elements into even the most competitive dispute settings.

Despite an often shared interest in finding a settlement, each group may wish to strike the bargain that most advances its interests. In any particular negotiation, this desire to advance the group's interest introduces a competitive element into the negotiations. Often, what one group or individual wins, another loses.[1] Thus, there will exist a tension between the desire to reach a settlement that is acceptable to the other party and the desire to advance one's own cause. No matter how much our husband and wife love to spend time together, the husband still wants to spend the time at the baseball game and the wife at the ballet. Within a country, laws, government action, and a shared community experience can help accentuate the cooperative aspects of any negotiation and the common interests that link the negotiating groups. In relations between nations, international organizations or individual mediators can act to accentuate the common interest between nations that are present in any dispute. Accentuating the cooperative elements of a particular dispute often provides the key to reaching a settlement.

Collective bargaining in industrial relations offers a negotiation setting in which the bargaining groups possess major common interests that unite them in their search for a negotiated settlement. Both sides share a fundamental interest in the profitability of a company or an industry. If a company goes bankrupt, workers will lose their jobs. Recent history illustrates the power of the common interest in avoiding bankruptcy. In 1979, the management of Chrysler and the United Auto Workers union lobbied Congress for loan guarantees to prevent the auto company's collapse. Both labor and management had a shared interest in survival of

[1]This event is often called a *zero-sum situation* and follows the game theory convention of giving a numerical score to the most favorable outcome of "1" and a score of "−1" to the worst. When one individual bargains to a settlement that gives him or her a positive, the other gets an identical but negative score. Together they add up to zero.

Chrysler. Similarly, when profits grow, workers can gain an increase in wages. Under duress, this usually tacit understanding can be institutionalized. In 1981, unions working for Pan Am airlines agreed to a 10% wage cut to prevent the bankruptcy of the airline. In return for this decrease in pay, Pan Am established a profit-sharing program that promised to reward workers if operations became profitable. In general, workers who toil in profitable industries earn higher wages than those in declining industries. Thus, workers have an interest not only in a firm's surviving but also in its flourishing.

Although the common interest of labor and management in the success of the firm lies at the heart of the bargaining relationship, it rarely surfaces in the public statements of the negotiators. As bargaining proceeds, the mutual benefits of agreement become buried beneath bargaining rhetoric. As the date of the contract's expiration approaches, the threat of a strike and the mutual damage that it will bring generates a common interest in avoiding the costs of a prolonged strike. In a strike, the company loses profits and the workers lose their wages. The failure to reach a settlement subjects both sides to heavy losses. Thus, the structure of the collective-bargaining relationship creates a common interest in avoiding this loss that links management and labor.

Labor and management, in addition to having common interests that spring from the economics of production, also share a common interest in workplace harmony. Labor and management must interact daily on the job. This ongoing relationship develops common human interests that are independent of many economic considerations and can support a cooperative resolution of bargaining differences.

As bargaining rhetoric makes quite clear, despite these common interests, labor and management have competing interests over the specific terms of the labor contract and the magnitude of the settlement. Higher wages, restrictive work rules, job security, and seniority provisions both limit a firm's profitability and managerial discretion. Thus, on the specific issues of contention, management and labor have directly competing interests. What one side wins, the other loses.

Law, however, accentuates the common and complementary interests in labor–management relations, limits the tactics of industrial conflict, and provides a binding endorsement of negotiation. A major factor influencing the negotiation of a labor contract is that the negotiators seldom have any alternative other than reaching a negotiated agreement. Once a union obtains formal recognition through an election in the workplace, law requires management and labor to bargain in good faith (National Labor Relations Act, 29 USC Section 158). After the formation of the union and the negotiation of the initial labor contract, both sides generally realize that they must reach an agreement whenever a contract expires. Law links labor and management together so strongly that a negotiated settlement generally offers the only practical resolution to differences. The almost permanent nature of this link creates a common interest in the firm's or the industry's long-term success.

Laws and court decisions act to limit the destructive consequences of the competitive aspects of industrial relations. Laws now regulate strike activities in ways that reduce the potential for violence. Sit-down strikes, mass picketing, and the use of violence are illegal strike activities (McNaughton and Lazar, 1954). Unlike in the early days of labor organizing when unions struggled to gain recognition, the consequences of industrial strife are measured in dollars, not dead bodies. The resumption of work quickly ends the economic losses.

In international relations, nations will not negotiate unless they can both realize some gain and negotiations offers the easiest way to obtain it. Iklé (1964) suggests that nations will negotiate only when there exist both common and conflicting interests: "Without common interests there is nothing to negotiate for, without conflicting interests there is nothing to negotiate about" (p. 2). In relations between the United States and the Soviet Union, the avoidance of nuclear war acts as a particularly strong common interest that cautions against confrontations that can readily escalate. In negotiations with friendly or allied nations, economic advancement, mutual protection, and stability often provide particular common interests that help negotiations reach settlements (Iklé, 1964; Neustadt 1970).

Within this broad framework of common interests, however, nations will have competing interests over the actual terms of a particular settlement. In arms control agreements, the United States and the Soviet Union invariably differ over how to classify, count, or compare the weapons of one country with the weapons of the other. Within NATO, the larger common interests that link the Western democracies do not prevent unending disputes over the choice of a particular weapon, where it will be manufactured, and who will pay for it.

Although diplomatic traditions and institutions such as the United Nations or the Organization of American States structure the conduct of international negotiations, no laws either require bargaining or limit the extent of destructive conflict. Nations will choose to negotiate only when bargaining offers the best alternative. In any conflict situation, nations can choose to file protests with international organizations, break off diplomatic relations, establish trade embargoes, negotiate, or use force (Iklé, 1964). Each alternative presents a country with a different set of potential consequences. Nations will choose to negotiate only when bargaining offers the best alternative. However, unlike many alternatives, the implementation of negotiations requires the agreement of all the conflicting parties. Force requires only unilateral action.

Negotiations over proposed development projects differ greatly from labor–management relations. The parties to a environmental-development conflict possess less clear common interests than do union and management. Although everyone shares the same ecosphere, views about the importance of any one activity can range widely. Utility managers and engineers, for example, may strongly believe that the heat generated by a power plant's operations causes no damage to the environment. They resent the large economic costs borne by the company and its customers when opponents succeed in shutting down plants. Opponents, on the other hand, may value the damage caused by thermal discharges quite highly, fail to see the need for a new plant, and insist that companies and the public pay large sums to preserve the sanctity of the environment. Environmental, industrial, and other viewpoints will range over all positions.

Unlike in collective bargaining, the collapse of environmental negotiations does not bring a consequence that both sides dislike or a new realization of common interests. In environmental conflicts, when negotiations fail to reach a settlement, parties can shift to judicial or regulatory forums that can render a final and binding decision. The effects of not reaching an agreement in a developmental conflict can be highly asymmetrical. Long court delays may have disastrous economic effects on a developer but may impose much smaller costs on litigants. In addition, courts offer environmental groups a familiar forum for resolving disputes. They understand how this forum works and have attained major successes in it.

Although the lack of common interests makes negotiations over development projects resemble international relations, they also lack the customary supporting structures and traditions that moderate international negotiations. Developmental-environmental negotiations take place over a much narrower set of issues than do the negotiations between nations. There is never an existing agreement that needs extension, nor can disputants normalize relations in any meaningful way. Disputants all live under the same system of laws, and they cannot create new institutional arrangements, except by changing the laws. Negotiations between conflicting groups can lead to a new configuration of a project's design, or to the abandonment of a particular project, but unless both sides will likely derive gains from bargaining that they would likely lose from litigation, bargaining will not take place. Most importantly, failure to resolve a dispute through negotiations does not lead to adverse consequences to the competing groups. Failure to negotiate in an environmental dispute brings litigation and delay, not war.

Neither law nor custom compels competing groups to negotiate an environmental or developmental conflict; rather, law insures a citizen's right to litigate. Developers have little incentive to negotiate if they believe that they can build a plant economically despite objections, or if they believe that no concessions will eliminate opposition by splinter or fringe groups. Opponents have little incentive to negotiate if they believe that they can win in a judicial forum. Likewise, if opponents of a technology adopt a long-run strategy that seeks to raise its costs through litigation and delay, then bargaining will hold no attraction for them.

Mediators can function effectively in all three settings. The task of the mediator is to accentuate the cooperative elements present in a dispute. The difficulty of this task depends on the legal framework or traditions that regulate a dispute. In labor–management relations, laws establish structures that decrease the potential for violence and increase the likelihood of a negotiated settlement. These laws reduce the tasks that mediators must generally perform. Despite this support of law, in first-time negotiations mediators can still assist the bargainers to establish a negotiating relationship that accentuates the cooperative elements of bargaining. Mediators can stress the importance of establishing a constructive bargaining relationship.

Even without the aid of laws to structure disputes, mediators can assist those who seek to negotiate by focusing on the common interests that tie the conflicting parties together. In international disputes, mediators can act to persuade the conflicting parties of the importance of the common interests that unite the countries. Young (1970) argues that mediation actions taken by the United Nations have facilitated the resolution of international crises peacefully through just such an accentuation of the common interests that both sides possess.

In environmental disputes, a common or complementary interest between the conflicting parties is not always present, or if present, not very strong. When it is present, a mediator who emphasizes this common interest can facilitate the resolution of a conflict through negotiations. In the dispute over the proposed construction of a dam on the Snoqualmie River, environmental groups feared that the dam would cause rapid development in the flood plain, while farmers in the flood plain feared the economic loss and danger of floods (Cormick and McCarthy, 1974). Mediators reminded the disputing farmers and environmental groups that they shared a common interest in seeking a solution that would limit flooding. Mediators stressed that if a flood caused economic hardship to the farmers or a loss of lives, then the community would blame the environmentalists who opposed the dam. This blame, they argued, would produce a serious reduction in the community's acceptance of the legitimacy of environmental concerns. After gaining a recognition of this common interest in flood control, the mediator was able to help both sides to develop a settlement in which environmentalists won development restrictions that would limit construction in the flood plain, while the farmers won protection from flooding. Thus, through emphasizing the common interests that unite the parties in a dispute, negotiations may succeed even when laws fail to support bargaining.

The Method for Recognizing Groups As Legitimate Participants in Negotiations

The mechanism by which groups recognize the rights of competing groups as legitimate participants in a dispute affects the chances of reaching a negotiated settlement. In labor–management relations, laws and government not only regulate group structures and leadership but also act to resolve issues over who represents workers and when collective bargaining must take place. In international relations, common practices lead most nations to recognize the governments of competing nations as having the authority to negotiate on behalf of their countries. In environmental-development conflicts, neither domestic law nor prior practice solves the question of who can legitimately represent a particular interest.

In a labor conflict, the major step leading to collective bargaining is the recognition by labor and management of the other party. The most prolonged and violent labor disputes in this country arose when management failed to acknowledge a union role in the determination of wages and working conditions. Even today, recognition fights, such as the textile workers' attempts to organize workers at J. P. Stevens or the United Farm Workers' efforts to organize California vineyards, can become bitter and violent. To reduce the potential for violence in industrial conflicts, law requires the National Labor Relations Board to establish routine election procedures and criteria that, if met, legitimize union representation and require collective bargaining on a labor contract. Management must bargain with union negotiators if over 50% of the workers in a bargaining unit vote to accept a particular union to represent their interests (National Labor Relations Act, 29 USC, Section 154).

Despite these clear rules, a special bitterness not associated with contract renewals often mars first-time bargaining. Neither side knows the other very well or has familiarity with collective-bargaining procedures. Expectations may diverge sharply from reality, which can lead to strikes, violence, and other tests of strength.

In international disputes, one individual or group generally holds the reins of domestic power. Sometimes, however, nations will refuse to recognize another power as the legitimate voice for a people or territory. The United States, for example, refused to recognize the People's Republic of China until 1979. Nevertheless, conflicting nations realize that short of achieving their ends through force, they cannot settle major conflicts without talking to representatives of the governments which they refuse to recognize.

Mediators can help disputing parties that do not officially recognize the legitimacy of a group's participation in a dispute by facilitating informal communications when the terms for formal face-to-face communications are themselves an object of contention. This technique has proved most successful in international relations, where a third country can facilitate communications that precede formal recognition. In talks preceding the U.S. recognition of the People's Republic of China, for example, Pakistan passed messages and arranged for secret exploratory talks before the formal visit of President Nixon to China (Kissinger, 1979).

In development conflicts, the identification and recognition of groups as negotiation participants poses serious philosophical and practical problems. Whom do the different intervening groups represent? Who should participate? These questions have no clear answer. In theory, governmental regulatory agencies have the mission of protecting the public interest while overseeing the construction of new facilities. In practice, there is no single public interest that can offer clear decision rules to administrators. On the other hand, since the tenets of U.S. political philosophy hold that the American government should represent the interests of the people rather than rule them, it is natural for government agencies and legislators to seek the views of special-interest groups. Many federal laws require public participation (in some form) in regulatory decisions and allow citizen suits to enforce this right.

In disputes resolved through negotiation and mediation, the recognition of groups opposing a particular development project generally has evolved as the dispute progressed. Those groups that can demonstrate through litigation their power to block or obstruct a project generally gain recognition as dispute participants. In the Colstrip Power Plant controversy, the Northern Cheyenne tribal council was both a strategic litigator and a natural representative of the tribe. Negotiation efforts generally succeed only by recognizing everyone. This all-embracing approach creates a complex multilateral bargaining situation in which agreement becomes unlikely.

Since law fails to establish any rules for recognizing legitimate bargaining participants in a development dispute, efforts to initiate negotiations have resembled those that attempt to bring a hostile nation to a bargaining table. In disputes over the construction of a dam on the Snoqualmie River in Washington State and over the construction of the White Flint shopping mall in suburban Maryland, mediators first elicited from the conflicting parties a formal recognition of the legitimacy of each party's claims and a commitment to negotiations. Government officials also endorsed the negotiation effort. Since environmental laws protect the right of almost any individual or group to litigate a government decision, in these successful negotiations the parties recognized virtually every individual and group as a legitimate participant in any formal negotiation effort.

Mediators have helped the process of initiating negotiations over development by providing a means of communication apart from the adversarial forum of the courts and by accomplishing the logistically complex task of arranging meetings. Without legislation patterned after labor law to routinize the recognition of legitimate participants in a development decision, the use of environmental negotiations will likely continue to require the work of mediators to facilitate contacts prior to the initiation of formal bargaining. Although the actions of mediators will never routinize recognition as clearly as labor law resolves questions of union recognition, mediator action can serve as an informal substitute for the lack of formal procedures. These actions, however, require the delicate diplomacy more common in mediating disputes between nations.

The Balance of Resources

Resolving a dispute through negotiations is much simpler when the competing groups possess relatively similar resources. Asymmetries of power tempt the more powerful groups into believing that they can accomplish their aims without negotiation or compromise. They may view bargaining as unnecessary. If this assessment is incorrect, then a destructive test of strength may ensue as opposing groups seek to demonstrate and use their power.

Great asymmetries of power also create psychological barriers to bargaining. In particular, the weaker individual or group may feel intimidated by negotiation. Bargaining requires that a bargainer possess the confidence that his or her assent brings an important element to the settlement. Rage or fear felt by groups with little power can preclude the communications necessary for a constructive resolution of a conflict (Deutsch, 1973).

The history of U.S. industrial relations shows the failure of collective bargaining when there existed great asymmetries of power. Before the 1930s, management possessed great strengths in its dealings with unions. Armed with antitrust legislation, security forces, and court injunctions, management possessed great ability to limit a union's power to organize for bargaining. Only skilled laborers held sufficient power to bargain. When skilled laborers struck, they withheld their services. Since the skilled character of their work prevented quick replacement, a collective refusal to work shut down a plant.

Unskilled workers faced great difficulties in their initial attempts to organize. Management could easily hire substitutes. Thus, effective strikes proved extremely difficult to mount, and early mass unions often failed. The Knights of Labor, a broad-based labor movement organized to advance the rights and wages of workers, illustrates the failure of this method of organizing. In 1878, it had 20,000 members; in 1887, 700,000; but by 1893, 75,000 (McNaughton and Lazar, 1954). The failure of this movement to win strikes created resistance within the AFL and its craft unions for any attempts to organize workers along industry lines.

Public policy of the 1930s radically changed the power of unions. The passage of the National Industrial Recovery Act and then the National Labor Relations Act (Wagner Act) not only put public policy behind collective bargaining but facilitated the organization of large-scale industrial unions. The rapid rise of the CIO, which faced resistance from the old-line trade unions, illustrates how effective this law was in promoting collective bargaining. Membership in the CIO increased from 3.6 million in 1936 to 17 million in 1950 (Herman and Kuhn, 1981). Both the establishment of large bargaining units by the National Labor Relations Board and the legal requirement to bargain with all unions winning a bargaining unit election created conditions conducive to the rise of industrial unions.

Although law sets the broad dimensions of management and union powers, many specific factors will affect the resources available to each side in a contract negotiation. High profits, high inventories, and little competition will strengthen management's bargaining hand. Strike funds and firm vulnerability to lost production strengthen the union's bargaining hand. General economic conditions affect the overall shape of a settlement. Corporate and union reports provide information on the resources each side can use in a strike. Although the balance of power can change, law establishes a fairly balanced power setting that facilitates bargaining.

In international relations, the relative equality or inequality in resources can affect the relations of one country with another. Kissinger (1982) states that Egypt–Israel peace negotiations became possible only when the 1973 war had restored the self-confidence of the Arabs and demonstrated to Israel that it could not rely on arms alone to maintain its security. In his view, only after achieving a relative parity between the two sides could both sides recognize a mutual interest in peace that could withstand the give-and-take of negotiations. Unlike in labor–management relations, the changes of attitudes and power necessary for negotiations followed not changes in law, but the waging of a war, the most destructive form of conflict resolution.

Disputes over development projects have traditionally involved government agencies and developers, typically rich in resources and political power, and local community or environmental groups, usually low in resources and uncertain of their power. Over the 1970s, major changes in legislation have enhanced both the power and the resources of local and environmental groups. The National Environmental Policy Act gave groups broad powers to intervene in opposition to proposed projects. The 1980 Equal Access to Justice Act enables groups that win an environmental case in court to receive an award of legal fees. Court decisions over the last decade have increased the ability of opposition groups to obtain legal standing in a dispute. Groups opposing government decisions now have the legal power and better access to resources to express their views. This access to resources, however, is available only to assist the representation of interests through litigation.

Mediators can help disputants in those conflicts in which imbalances of power dominate the perceptions of the disputants. If the perceptions of disputing groups do not truly reflect underlying realities, then a mediator may help the disputing parties to alter their perception and to avoid unnecessary tests of strength or destructive conflicts (Jackson, 1952; Pruitt, 1971). The ability of a mediator to debunk false impressions depends critically on whether both sides trust the mediator, whether the mediator's perceptions are accurate, and whether the disputing groups will accept new information (particularly bad news). The openness of groups to new information will change over time. Just after the start of a destructive conflict, both sides may become more open to a reassessment of their judgments. When there exist real asymmetries of power, then an important function of a mediator is to help prevent the development of rage or anger in the group with little power (Deutsch, 1973). Sometimes, the mere presence of a third party can lead to more constructive norms of interaction. These mediator actions can help maintain the rationality necessary for meaningful negotiations and reasoned responses.

The Periodicity of Negotiations

One important feature of any negotiations is whether bargaining is a one-time affair or part of an ongoing relationship that requires frequent and regular negotiating. When two groups bargain repeatedly, each side develops a familiarity with the procedures of collective bargaining and with the characteristics of the individual negotiators (Douglas, 1957). This experience can prove invaluable as both sides learn to differentiate between public postures, firm commitments, and negotiable issues. Both sides learn a vocabulary of signals that carries a meaning unique to their negotiations.

Repeated negotiations also diminish the advantages of attempting to gain total victory. If such attempts succeed, the alienation that they produce can damage future relations. Since the relative balance of power of the two parties may shift over time (Simkin, 1971), a prudent long-run strategy for both sides may require the generation of goodwill between the groups.

Labor and management establish continuing relations with each other. Once a union is formed and collective bargaining begins, the union and management can expect to negotiate whenever a contract expiration date nears. This ongoing bargaining relationship allows each side to gain a great familiarity with the communication and bargaining style of the other. Only when union or management leadership changes can one expect any communication problems. The continuing relationship between labor and management permits both sides to take a longer view. When business suffers a downturn, or the fortunes of a particular firm sour, unions may accept a leaner contract with a tacit understanding of larger settlements when conditions prove more favorable.

International negotiations can take place either as part of a series of negotiations between the same parties concerning the same subjects, or as a one-time effort to resolve a particular crisis. Airline landing agreements, for example, are periodically renegotiated between countries. The financing of NATO troops and decisions on the size and placement of forces receive periodic review and renegotiation. Once again, when such an ongoing relationship exists, the likelihood of successfully negotiating a resolution of differences increases. Every negotiation round provides the bargainers with a familiarity with the negotiation process and the style of negotiations. This can facilitate communications between the parties and decrease the chance of a misunderstanding or bargaining error. In addition, prior agreements anchor the expectations of the negotiating parties to a narrow range within which smaller remaining differences are easier to settle. When each side fulfills the conditions of the prior agreement, trust can build between the parties as each side forms expectations that the opposing side will honor its agreements (Iklé, 1964). This trust can facilitate the course of negotiations (Deutsch, 1973).

In an international crisis, the conflicting nations seldom possess a bargaining relationship that will facilitate the resolution of differences through negotiation. Initiating discussions will require a major break with the past. For example, following the 1973 war between Egypt and Israel, the countries lacked any formal diplomatic relations. Establishing diplomatic contacts, a requirement for negotiations, necessitated a major new initiative and the assumption of a major risk. In the 1973 war, after making major military gains, Sadat expressed a willingness to begin a negotiation with Israel. One major complication stood in the way of negotiations: Israel had violated a ceasefire to entrap Egypt's Third Army in the desert (Kissinger, 1982). Despite this situation and against the counsel of his advisers, Sadat agreed to let the army remain a hostage to Israel while Kissinger attempted to initiate a peace process.

In Kissinger's view (1982), Egypt had to think of peace with Israel as a psychological rather than a diplomatic or tactical problem. Risking the army for peace would help Egypt to change Israeli perceptions. Following Sadat's negotiation initiative, Kissinger's diplomacy led to a conference in Geneva on December 21, 1973, that brought Egypt and Israel face-to-face in negotiations for the first time. This began the long process that produced the Sinai accord in 1974, which set a ceasefire and disengagement of troops; a second Sinai treaty in 1975, which set a withdrawal of Israel to strategic passes in the Sinai; and the Camp David Peace Treaty of 1979, which led to Israel's withdrawal from the Sinai and the establishment of diplomatic relations between the two nations.

This particular sequence of negotiations also illustrates the advantage of negotiating a resolution of a conflict through a series of bargaining sessions and intermediate treaties. Repeated bargaining helps each side to develop an understanding of the bargaining styles and positions of its opponent and creates trust that can produce more fruitful future negotiations. Professional diplomats will often take a crisis situation and divide a problem into a series of smaller problems that permit sequential resolution. When the two sides desire a negotiated resolution of differences, this technique can assist countries to develop a relationship that makes settlement possible.

Environmental-development bargaining generally requires that the disputing groups meet for the first and only time in an unfamiliar setting. The course that the negotiations will take is never clear. The ambiguities of many environmental laws introduce real uncertainties into each side's assessment of its own rights and the rights of its opponents. Since negotiators need not deal with each other in the future, the temptation to strive to win an extreme settlement may prove irresistible. These uncertainties of law and of the negotiation context may cause each side to have trouble understanding what the other wants or means.

When bargainers lack a history of bargaining, an orientation provided by a neutral outside party, such as a mediator, can supply the bargainers with an understanding of how negotiations will develop. This can reduce some of the uncertainties inherent in first-time negotiations and help set mutual expectations in ways that promote settlements. The fact that negotiation over development disputes is often called *environmental mediation* doubtlessly reflects the large role played by the mediator in initiating and managing these negotiations.

Generally the mediator's bargaining tasks are simplest when the disputing groups negotiate frequently in an ongoing relationship.[2] When the conflicting parties have a history of resolving issues through negotiations, the likelihood of bargaining success increases for several reasons. Past experience facilitates bargaining communications and reduces the possibility of misunderstandings. Prior agreements on related issues form the expectations of both sides about what the final agreement should look like. Thus, negotiators are less likely to face the shock of learning that their demands are unrealistic, and less likely to commit their organizations to untenable and inflexible positions. A long-term bargaining relationship may allow negotiators to defer thornier issues to the future, when either the momentum of history or good fortune may alter the issues in ways that make them more amenable to resolution.

In situations where a prior or continuing bargaining relationship fails to exist between conflicting groups, a mediator can help the negotiators to develop a bargaining process that facilitates the search for an agreement. The key actions that a mediator can take arise from the special role that he or she plays in managing bargaining communications. In new or one-time bargaining situations, a mediator's intervention may help to dissipate some of the mistrust and fear that can stifle communi-

---

[2]Once again, exceptions exist. Hatred or animosity may prevent settlements even when reason suggests common interests. In labor relations, International Harvester is known for its bitter union–management relations, which have helped push it toward bankruptcy.

cations. The negotiating setting becomes trilateral, and the mediator serves to moderate bargaining interaction.

If a mediator has more technical skills at bargaining than the negotiating parties, he or she may suggest ways to split a crisis or problem into a series of smaller problems. With a mediator's guidance, the conflicting parties may establish a bargaining process in which bargaining successes and kept promises at the early stages create a momentum that facilitates progress and develops trust that enables each disputing group to take larger risks in the search for a settlement (Jackson, 1952). In particular, solving simple problems first can allow the bargainers to resolve misunderstandings that arise from either bargaining or communications styles before they address major issues of contention. In the negotiations between Egypt and Israel after the 1973 war, the nations deferred consideration of such thorny issues as the status of Jerusalem and the West Bank Arabs. Although these issues remain difficult to resolve, the series of successful agreements between Egypt and Israel make a peaceful resolution of even these issues not unthinkable.

When conflicting countries cannot defer the resolution of a principal issue (such as in a hostage seizure, where the status quo of continued imprisonment is unacceptable), then the prospects of successful division into smaller issues are dim. When establishing trust is not a problem, however, the simultaneous consideration of several issues can increase the chance of overall bargaining success (Pruitt, 1981) because it permits mutually beneficial concessions.

The Number of Bargaining Groups

The number of groups participating in negotiations affects the likelihood of achieving a settlement (Raiffa, 1982). Bilateral negotiations offer a much simpler bargaining context than multilateral negotiations. In bilateral negotiations, only two sides must agree. Multilateral negotiations generally require a more difficult search for consensus and alter the negotiation process itself. Negotiations may continue as a series of coalitions form and dissolve. All else being equal, negotiation structures that facilitate bilateral bargaining offer a situation that simplifies the search for a negotiated settlement.

In the United States, laws establish bilateral negotiation in collective bargaining. The Taft-Hartley Act requires the National Labor Relations Board to certify the winner of jurisdictional labor elections as the sole bargaining agent for a group of workers. Management bargains with one union that represents all workers in a particular bargaining unit. This bilateral structure—management bargaining with a single union—is not universal. Laws regulating the coal-mining industry in France, Germany, and Italy do not require exclusive representation, and national agreements are signed by more than one union (Dunlop, 1958). These other arrangements complicate the task of bargaining.

International relations can be either bilateral or multilateral. Once again, when negotiations are multilateral, bargaining can involve a set of shifting coalitions of nations looking for a consensus. In the Eighteen Nation Committee on Disarmament meetings between 1962 and 1969, the participating nations formed bargaining coalitions as part of the search for a disarmament agreement. The 17 participating states broke down into three coalitions: the West, the East, and the nonaligned nations. Thus, multilateral bargaining was reduced to a loosely structured trilateral situation based on commonality of interests (Stenelo, 1972). This simplification offered a more manageable structure for resolving bargaining issues.

Conflicts between developers and groups opposing their projects can require multilateral negotiations. Each participating group has its own interests, which it seeks to promote. In the controversy over the construction of two power plants at Colstrip, Montana, the major groups included a power company, the EPA, a state regulatory agency, a group of ranchers, and an Indian tribe. A negotiated settlement was reached only after the EPA dropped out of the dispute and Montana Power and the Northern Cheyenne ignored the ranchers and created a bilateral situation.

When government regulatory groups have discretionary power, they can alter their role to support negotiations and assist bargainers to reach a settlement. In the controversy at Colstrip, the Montana State Department of Environmental Conservation, which played a small role in the dispute, required the Cheyenne tribe and Montana Power to negotiate a plan to control the impacts of construction on the local communities. This formal endorsement of negotiations helped legitimate the search for a bargaining settlement.

Deadlines

Deadlines serve as effective prods to those motivated to negotiate a resolution to a particular conflict (Pruitt, 1981). When a bargaining relationship has certain deadlines, experienced negotiators and mediators can use them to spur bargaining progress. If bargaining sessions lack natural deadlines, then mediators or the negotiators themselves may create deadlines to structure bargaining.

In collective bargaining, both sides race against the deadline of contract expiration and a likely strike. Although the contract could be extended by a simple agreement between both sides, this rarely occurs. Neither side wishes to see the effectiveness of the contract expiration date reduced. The losses that both sides face when a strike occurs can greatly aid in the making of a settlement. In the late stages of negotiations, the differences between the two parties are often less than the costs they will bear if they fail to reach an agreement. Many disputes are resolved in the waning hours of the contract. The contract expiration date (and fear of strike costs) helps both sides bridge the bargaining distance that remains between them.

International negotiations generally lack action-forcing deadlines. Negotiations can often drag on endlessly. Nevertheless, internal political pressures, such as upcoming elections, or a prearranged deadline, such as a future summit meeting, can create pressures for reaching a timely settlement. Kissinger, in *White House Years* (1979), recalls the frantic efforts of the United States and China to negotiate the wording of the Shanghai Communiqué. Of particular concern was the wording used to describe the positions of the United States and China on Taiwan. Kissinger says that "each side pushed the other other against the time limit [generated by the visit schedule] to test whose resiliency was greater" (p. 1308). In this situation, the timetable of Nixon's visit set a natural deadline that required some statement of the positions of the United States and China. Both sides used this deadline to facilitate bargaining progress and bureaucratic decision-making as they sought to avoid the disappointment of a public failure to reach some major resolution of the issues separating both sides.

At other times, negotiations can drag on endlessly. Negotiation to end the Vietnam War began in in 1967 and did not produce an agreement until January 1972. Kissinger (1979) argues that bargaining progress became possible only when the 1972 U.S. presidential election presented Hanoi with a coming deadline. After the likely reelection of Nixon, Hanoi would face a hawk with a mandate for "four more years." In July of the election year, serious negotiations with the North Vietnamese commenced and produced a breakthrough in October, prior to the November elections (Kissinger, 1979). In Kissinger's view, the internal politics and election cycle of the United States created pressures on the Vietnamese for a timely settlement. Kissinger claims that had McGovern mounted a stronger campaign, it is likely that the Nixon government would have felt deadline pressures to settle before the upcoming election, while Hanoi would have desired to wait for the new administration.

Environmental-development conflicts seldom generate effective deadlines. If negotiations fail, then both sides resort to litigation and the slow pace of legal movement. There is seldom a major incentive to close the remaining distance between the opposing sides since, unlike labor negotiators, they do not face any disaster from the failure to reach an agreement.

In many environmental-development disputes, the costs of failing to agree are highly asymmetrical. If a developer has sunk large amounts of money into a project, then delays cause the firm to bear interest charges. Those producing the delay bear only the court costs necessary to win the construction stoppage. When one party opposes a facility per se, then the delay of construction is a partial victory. When opponents seek to stop construction, stopping it for many months is one step toward stopping it forever.

The creation of deadlines can help spur bargaining progress if a failure to reach a settlement places burdens on those bargaining (Pruitt, 1981). In the dispute over the construction of a dam on the Snoqualmie River in Washington State, Governor Evans supported an effort to negotiate a compromise between environmental and agricultural interests. Governor Evans helped structure these negotiations by setting deadlines on the attempts to mediate the dispute (Cormick and Patton, 1977). He threatened that if the negotiating parties failed to reach an agreement, he would resolve the dispute through unilateral actions. The negotiators successfully used these deadlines and, despite months of drift, generated a settlement.

When negotiations lack a formal deadline structure, a mediator may help in the bargaining sessions by giving important dates the force of deadlines. In international negotiations, for example, a mediator may use national or religious holidays as deadlines for spurring negotiation progress. By giving these prominent dates added significance, a mediator or negotiator can help create some deadlines to structure an open-ended bargaining situation and spur bargaining progress. Once again, it is important that both sides prefer to reach an agreement so that the failure to meet a deadline causes both sides to regret the failure to achieve a bargaining settlement.

Designing Binding Agreements

The ability of negotiators to make binding agreements is important to the success of negotiations. Once again, this feature of the negotiating setting varies greatly. Labor–management disputes are concluded by legally enforceable contracts. International negotiations end with formal treaties. Negotiators in environmental-development conflicts, however, face difficulties in finding instruments that can legally bind parties to a settlement.

In those conflict situations that are traditionally concluded by a formal agreement, the mediator can help the negotiators to focus on the terms of the final settlement. Where the negotiations lack a traditional instrument that formally specifies the terms of an agreement, then the mediator can facilitate a settlement by assisting the conflicting parties in designing an instrument that can either legally bind parties to a settlement or create a multistep process that encourages compliance and generates trust between the groups.

A settlement offer ratified by the union creates a legally binding and enforceable contract. Theoretically, the enforcement of the terms of the agreement lies outside the framework of negotiations. Either side may appeal departures from the contract's provisions to a court of law. Each side unequivocally binds itself to act according to the terms of the agreement. In practice, the ongoing relationship of the union and management in the production enterprise provides a strong incentive for each side to keep its word. Contract violations can lead to strikes or bitter labor–management relations. In addition to the terms of agreement, most contracts develop grievance procedures to settle disputes without strikes or lockouts.

Successful negotiations between sovereign nations usually lead to a treaty that formalizes the terms of agreement. Unlike in contracts between individuals in a country, there exists no higher power than can enforce the terms of a treaty. Iklé (1964) suggests that nations honor treaties for three reasons. First, when the terms of a treaty codify reciprocal actions, it usually remains in the interests of both parties to respect the treaty. Airplane landing agreements often possess this reciprocal nature. For example, a nation permits the planes of another country to land as long as that nation allows its planes to land.

Second, when a nation signs a treaty, it gives a formal promise to support some action. Since a country has a continuing existence, its violations of a treaty will affect its relations with other countries and may curtail its ability to make new treaties. Likewise, a country may take retaliatory actions against a nation that violates a treaty with it to insure that other countries will meet their commitments. Violations of treaties may result in retaliatory steps that would have appeared unjustified or foolish without a treaty. As a formal document, the treaty itself can both make violations of its terms unattractive and increase the rewards of honoring them.

Third, treaties can affect the domestic environments of signatory nations. Treaties may create internal bureaucracies with a mission to fulfill treaty terms. These groups form an internal force to maintain policies and practices. In the United States, the violation of a treaty may prove politically unwise even if strategically prudent. The mobilization of legislative support necessary to meet constitutional requirements for legitimating new action or undoing the old treaty may prove difficult. Bureaucratic and legislative inertia may force a country to continue in a course determined by the treaty. Thus, although no court or higher authority exists to enforce the terms of a treaty, the existence of a treaty can create incentives that increase the probability that a country will honor it.

At this time, the ability of environmental groups to bind themselves to an agreement has severe limitations. To permit effective bargaining, negotiators must possess the ability to make commitments and promises to each other. In a development dispute, no one group can represent all environmental concerns or bind its membership to a course of action. For example, in an environmental-development dispute, opposition groups can sign a consent decree withdrawing their legal challenges to the issuance of a construction permit. They cannot, however, limit the intervention of other environmental groups in the proceedings on either the same or different grounds; nor can they prevent splinter groups from their own organization from continuing their opposition. When government agencies are participants, laws may limit their ability to bind themselves to any one course of action (Kretzmer, 1979).

The agreement that settles an environmental dispute often lacks simple procedures to settle disagreements that may follow the successful negotiation of a settlement. Unlike the production enterprise that relies on the daily cooperation of labor and management, the construction process has a definite completion. Once opposition abates, builders may revert to old plans. Without a mechanism to enforce the settlement terms, opposition groups may face a renewed struggle. On the other hand, once construction starts, a developer is even more vulnerable to delays caused by litigation. Large construction endeavors can remain vulnerable to relatively minor disagreements. Unless opposition groups and developers can design an enforcement and grievance mechanism short of litigation, developers may decide that the best policy is to wait for a final judicial resolution of issues before making a major commitment to construction. If the groups are unable to work out simple enforcement procedures, negotiations may offer conflicting groups few potential benefits.

A concern for a group's reputation can sometimes provide support for negotiated agreements where laws do not exist. In international relations, the concern for a government's international reputation arises from the continued existence of the country and its government. Unfortunately, in environmental-development conflicts, the disputing groups sometimes exist only for the duration of the conflict. Thus, they may have no formal reputation at stake in the bargaining process, and informal sanctions against agreement violations may not be strong.

Despite the lack of one formal mechanism for ending disputes in legally binding ways, bargainers have developed creative ways of committing their agency or group to a particular course of action. In the negotiated settlement preceding the conversion of the Brayton Point Power Plant to coal (Burgess and Smith, 1984), the Massachusetts Department of Environmental Quality signaled its commitment by incorporating terms of the agreement in the state's regulatory codes. Although the state still retained the power to amend the regulations, further changes would necessitate that the state justify its actions and would give the power company great legal opportunities to challenge such actions. In other situations, consent decrees signed by the disputing parties formally conclude disputes, but their legal status remains questionable.

Mediators can assist in developing settlements. When formal documents conclude negotiations, a mediator can help to focus discussions on the terms of the final agreements. As negotiations proceed, he or she can direct attention to the preparation of the settlement documents. When no document exists that can effectively guarantee performance, the mediator may suggest treaties or agreements that are implemented over time. If each action by one party is met with a reciprocal action by the opposing group, then a series of small steps implementing an agreement may prove acceptable even when neither side would agree to adopt a final settlement that required complete implementation at once. International treaties, such as the Egyptian-Israeli Peace Treaty of 1979, illustrate how a series of small steps implementing an agreement can prove more acceptable to all than large changes in the status quo. Over time, a treaty can become self-enforcing as each country becomes more committed to the process established by treaty. Similarly, in disputes over development projects, the difficulty of designing legally binding agreements complicates the task of achieving a negotiated settlement. As in international negotiations, a mediator may assist the two parties to develop an agreement that proceeds through a series of incremental and reciprocal steps that facilitates the implementation of the entire agreement.

Conclusion

Table 3 summarizes the features of a negotiation setting that affect the tasks that negotiators face. A quick perusal suggests that efforts to negotiate resolutions of disputes in international relations and development conflicts get little support from their environment. In industrial relations, laws have accentuated the cooperative aspects of labor–management relations, created rules for recognizing bargaining participants, helped strike a balance of power between union and management, created frequent negotiations, and limited the number of bargaining groups. Contracts both create bargaining deadlines and formally con-

Table 3
Bargaining Factors Affecting Negotiation

	U.S. industrial relations	International relations	Development disputes
Cooperative elements	Mutual interests in firm's profitability. Mutual interest in workplace harmony. Laws accentuate common interests. Laws regulate destructive tactics to limit the potential for violence.	Depends on dispute.	Rarely do common interests link developer and opposition groups.
Competitive elements	Workers desire high wages. Management prefers low costs.	Depends on dispute.	The polar forms of most disputes creates a we/they competitive environment.
Recognition procedures	Procedures set by law.	Formal diplomatic procedure prevails.	No set rules. Usually, developers meet with litigants.
Power balance	Depends on economic and organizational factors, but relatively symmetrical.	Vast differences often occur.	Usually asymmetrical. Opposition groups are often new and low on resources, but strong on legal powers.
Frequency of negotiations	Regularly.	Depends on relationship between countries.	Usually takes place for the first and only time during the dispute.
Numbers of groups	Bilateral relationship.	Bilateral or multilateral.	Often multilateral.
Deadlines	Set by contract.	Usually none.	Usually none.
Formal agreements	Labor contracts.	Treaties.	No routine mechanisms, but sometimes, participants develop an ad hoc arrangement.

clude negotiations. In international relations, negotiation is only one of a variety of policy alternatives. Unlike in labor–management negotiations, no laws necessitate bargaining, simplify recognition, limit conflict, or help strike a power balance. Diplomatic custom and international law have, however, structured bargaining between nations. Nevertheless, negotiations never need occur. Deadlines that force bargaining progress are often nonexistent. Treaties formally conclude international bargaining but there is no authority that will enforce treaty terms.

Once again, many of the features of current ad hoc methods used to resolve environmental-development disputes share many features with international relations. Almost no formal structures exist to facilitate bargaining. Just as a use of force can achieve goals in international relations, litigation often offers an attractive way for groups opposing a development project to make their case. Determining who should bargain is difficult. Opposition groups and developers often differ greatly in their power and resources. Negotiations seldom take place more than once, and there is no limit to the number of groups or individuals representing any one interest or concern. Deadlines rarely exist and there is no predetermined document that can routinely bind communities or groups to a particular course of action. Nonparticipants can freely raise issues that negotiators spent long hours deciding, and judicial review can unravel a negotiated package strand by strand.

An examination of the bargaining features in the negotiation setting allows us to make some general conclusions concerning those factors that are important for reaching a negotiated settlement:

- A bargaining process that accentuates the cooperative interest present in a dispute setting can facilitate the search for a settlement.
- When law or custom routinizes the process of recognizing groups as legitimate bargaining participants, then they can enable all to avoid a needless destructive conflict over the selection of bargaining participants.
- Environments that promote the development of an ongoing relationship between the disputing groups with periodic renegotiation can facilitate the search for a cooperative bargaining settlement. One-time negotiations can tempt bargainers to adopt competitive tactics that risk failure while searching for a big win.
- Bargaining settings that limit the number of negotiating groups simplify a search for a negotiated settlement.
- Bargaining deadlines encourage negotiating progress.

- Agreements that have the power to bind parties can help surmount obstacles caused by mistrust.

Mediators can respond to the structure of a negotiating setting in a variety of ways that can facilitate bargaining progress:

- When negotiators are bargaining for the first time, mediators can facilitate communications and help the bargainers to develop cooperative patterns of interaction.
- In any given conflict situation, mediators can accentuate the cooperative aspects of a problem. Often, a mediator can stress the benefits that a negotiated settlement will offer both parties.
- When laws fail to establish procedures that enable groups to recognize others as participants in a dispute, mediators, through contacts with disputants, can lay the groundwork necessary for bargaining. In international relations, a third nation can help adversaries to renew or establish formal relations. In development disputes, mediators, through discussion with disputants, may establish rules for the conduct of bargaining.
- When negotiations occur only once, or for the first time in a long-term relation, mediators can help each side to overcome the misunderstandings that bargaining will likely generate. When negotiations take place to resolve a particular, nonrecurring problem, mediators may suggest ways to divide it into several issues that allow sequential resolution through a series of negotiations and agreements implemented in steps.
- When the number of groups negotiating becomes large, bargaining can become unwieldy. Mediators can suggest ways of consolidating groups with common interests into a single bargaining coalition.
- When negotiations lack any deadlines that will conclusively end bargaining or will induce bargaining progress, mediators may create negotiating deadlines by adding to the prominence of particular dates. In international negotiations, elections and holidays can act as deadlines for either initiating negotiations or concluding agreements.
- When no document exists that can conclude bargaining in a binding or formal way, mediators may help bargainers to design agreement documents that enable each side to commit itself to the settlement terms.

These interventions can enable mediators to help negotiators to resolve disputes when the bargaining structures of a conflict situation fail to support bargaining. Despite the discouraging assessment that our comparison produces, the prospects need not remain so poor for negotiations over development disputes. To encourage a wider use of negotiations to resolve disputes over development projects, laws could provide these disputes with some of the structures present in industrial relations. Potentially, negotiations could offer both developers and opponents the possibility of joint gains. Developers would benefit from quicker and more predictable decisions that would be less likely to be challenged in court. It seems silly to have developed a siting process that necessitates the expenditure of millions of dollars before getting a final decision to construct a project. When opposition to a facility has a concrete cause, then mitigation measures or compensation may permit developers to gain community acceptance. Each side would then have an interest in resolving the dispute through negotiations.

Laws that facilitate negotiation may also change the climate for litigation. When citizens can voice their grievances in direct bargaining with facility developers, this explicit public participation in decision making may reduce the grounds for judicial reversal of the final decisions. If, as observers suggest (Stewart, 1975), courts are developing an interpretation of administrative law as interest representation, then such procedures that permit forceful and direct citizen participation may gain the endorsement of reviewing courts. Negotiation efforts may convince courts not only that an agency has acted to meet the procedural requirements of laws demanding public participation, but that the agency has given affected groups an opportunity to voice concerns and affect the decision.

As long as laws tie citizens' power and resources to affect public development decisions to litigation, it will be extremely difficult to channel disputes into a negotiating setting. Courts offer a good forum for altering asymmetries of power and resources and for protecting the rights of weaker groups. As new laws attempt to consolidate the rights won through litigation, legislators should consider whether tying disputes to adversarial forums offers a productive way of resolving a dispute. Tying rights and resources to adjudication almost insures that both sides will settle their dispute through a process that imposes large delay costs and legal fees on the developers of any major project and the users of its services.

CHAPTER 5

Tactical Aspects of Bargaining and Mediation

Introduction

In any bargaining situation, the representatives of the bargaining groups will take actions to enhance the likelihood that a settlement will favor their interests. Each side will try to determine how the choice of a particular issue agenda or bargaining tactic will affect the final settlement. Although bargainers choose issue agendas, strategies, and tactics to advance their interests, at times these actions may actually limit the chances of reaching a settlement. Even when a settlement could leave both sides better off than the continuation of the dispute, bargaining tactics can create obstacles that prevent a successful search for an agreement.

Although law and custom can help structure issue agendas and set limits on the bounds of acceptable negotiation tactics, negotiators can still fall into typical bargaining traps. This can prove especially true of attempts to resolve development disputes, where citizen groups often lack skills in bargaining. The participation of a skilled mediator in negotiation can help the negotiators to avoid the traps created by bargaining tactics or limited-issue agendas.

This chapter considers:

- Single-issue bargaining agendas;
- Multiple-issue bargaining agendas;
- Bargaining commitments and deadlocks;
- Discrete bargaining issues.

An analysis can help those establishing negotiations to avoid common pitfalls and can suggest ways of overcoming bargaining obstacles encountered. These obstacles are common to virtually all forms of bargaining. For this reason, this chapter draws less from an analysis of bargaining contexts than from the writings of bargaining strategists.

Single-Issue Bargaining Agendas

In the simplest bargaining situation, two negotiators seek to strike an agreement on the exchange price of a particular item. The two bargainers need not reach a settlement but can simply continue to search for another buyer or seller. The negotiations usually take place only once, and the bargainers have no relationship other than that which arises in the negotiations. The bargainers will each have a reservation price that any settlement must meet. For the seller, the reservation price is the lowest amount of money that he or she would accept for the item. For the buyer, the reservation price is the most money that he or she would pay. Through negotiations, the bargainers seek to determine whether their preferences, which their reservation prices reflect, will permit them to arrive at a deal that leaves them both better off.[1]

Bargaining over the price of a used car offers a good example of a simple form of single-issue bargaining. Often, individuals meet for the first and only time as a result of a newspaper advertisement that offers a car. If the bargainers cannot agree on an exchange price, each goes his or her separate way. Each bargainer will have a reservation price: the car buyer, a highest offer that he or she will make; the seller, a lowest price that he or she will accept. In California, these reservation prices are greatly influenced by published information. Each month, the *Kelley Blue Book* lists the prices at which dealers typically buy or sell used cars. When two individuals bargain over the selling price of a used car, these published prices often determine the range within which a bargain can be struck. No informed seller will offer more than the *Kelley Blue Book* top price, and no informed buyer will accept less than the *Kelley Blue Book* low price. A bargain can be struck whenever the seller is willing to accept less than the highest offering price of the buyer, or, more technically, when the reservation level of the seller is below the reservation level of the buyer.

Will the negotiators reach a settlement when an agreement is possible? What price will be the exchange price? It is impossible to tell. The buyer will always desire a slightly lower price, and the seller a higher

---

[1]See Raiffa (1982) for a technical discussion of these issues; see Cohen (1980) for a more familiar treatment of single-issue bargaining.

one. Although many exchange prices will leave both individuals better off than failing to close the deal, each would prefer a different price that leaves him or her a bit better off than any of the proposed possible settlements. Most exchange prices between the reservation levels of the seller and the buyer look pretty similar, and no particular price may readily recommend itself as the appropriate agreement price for making a used-car deal. What one bargainer wins, the other loses. In the dynamic tension of a search for a slightly better price, the deal itself may even fall through.

This dilemma, however, does not lead to the collapse of the used-car market. Deals are made every day. In our particular example, Californians will typically acknowledge the *Blue Book* prices and the inherent indeterminacy of the situation, and buyer and seller will come to an agreement by splitting the difference between *Blue Book* high and low price. Thus, the 50–50 split easily recommends itself to the negotiators (Schelling, 1960) as the one point different from all the others.

A dispassionate onlooker would likely see the pattern just described unfold over and over in many car transactions. The human reality of used-car buying, however, lies outside this analytical dimension. If you are buying a car, you are not bargaining in the abstract. It's your money, and it will be your car. If you are like me, it's probably much of the money that you have (and perhaps more than you have). Once you spend it, it's gone. Furthermore, cars offer an almost unlimited potential for absorbing additional cash. Even if you don't know much about cars, you know that they are expensive to fix. If you buy from an individual, you will never know as much about the car as he or she. If you buy from a dealer, you won't know as much as he or she does, but neither of you may know very much. You realize quite quickly that you are gambling with more money than you care to think about. Buying information (taking the car to a diagnostic clinic) is expensive (usually between $25 and $30) and a hassle. Even if the underlying strategic structure is exactly that portrayed above, it is overlaid with layers of emotions and anxiety.

Buying from a used-car dealer brings you up against a bargaining professional. Even if you research prices and performance through *Consumer Reports* or the *Kelley Blue Book,* you still will have little information. Lately, inflation has made the prices in *Consumer Reports* obsolete within weeks after the yearly issue that reviews used cars hits the stand. If you buy cars as old as I do, the *Blue Books* in the public library stop quoting prices. Consumer books, although marketed as helpful, just tell horror stories and scare one to death. Besides that, the salesperson at a lot shows you so many cars that all the information just swirls in your head. What does one do? Forget the research. Buy from the dealer with the honest reputation and the warranty. You may not get a good price, but you can limit your downside risks.

Even this most rational of activities, buying a car, quickly moves into a less rational realm where doubts, uncertainty, feelings of powerlessness, and potential regret dominate. I can only imagine what buying a family car is like. The potential for regret and recrimination must be boundless.

A variety of self-help books (Cohen, 1980; Fisher and Ury, 1981; Karrass, 1974) offer advice on how one can directly overcome the inhibitions and problems posed by these asymmetries of information and skill that normally arise in bargaining situations. They attempt to give cheery advice describing what you can do to prepare for those bargaining situations that inevitably arise. Perhaps most importantly, they offer generally sound advice and help one avoid the trauma that can cause one to flinch from bargaining. In a graduate seminar that I teach on dispute resolution, students report that after a few weeks they actively seek opportunities to bargain with merchants. They convince themselves that the worst a merchant can do is refuse to bargain.

In these situations where agreements are possible but indeterminacy or anxiety precludes a settlement, a mediator can sometimes help bargainers. Strategically, the involvement of a mediator may offer a good mechanism for allowing the bargaining participants to split the difference between their reservation levels. If both bargainers trust the mediator, then they can each tell him or her a price that represents their lowest or highest offer. The mediator can examine the prices submitted by the seller and the buyer and announce whether a deal is possible and what the exchange price is. This process saves time, energy, and anxiety. Psychologically, the intervention of a third party may help the bargainers by preventing fear from traumatizing those with less bargaining power. A sensitive mediator (or professional bargaining agent) can help the novice bargainers to avoid paralysis caused by the emotional content of the bargaining situation (Walton, 1969; Burton, 1969).

Although the 50–50 split helps solve many negotiations, some situations do not permit such an obvious resolution. Individuals bargain not only over prices, but over issues that do not permit continuous division or 50–50 splits. In those instances, no common numerator such as price will determine a range of possible settlement positions, and no single compromise position will have the prominence of a 50–50 division. If a group is divided between meeting in Boston or San Francisco, how can we measure their preferences? Furthermore, what represents a compromise position? Chicago may split the geographic distance between the disputing members, but will it prove a true compromise position? Those advocating Boston might prefer San Francisco to Chicago, while those advocating San Francisco may prefer Boston to Chicago. Even if Chicago does represent an acceptable intermediate position, Chicago will not likely prove an obvious compromise. Why not New Orleans, Minneapolis, Kansas City, or Dallas?

Mediators can prove particularly helpful in those disputes in which a large number of potentially acceptable solutions are available to the disputing parties, but the negotiators are unable to decide on a particular one (Schelling, 1960). In these situations, the mediator may, with the acceptance of both parties, recommend a particular settlement as one way of ending a dispute (Simkin, 1971). Both parties, although preferring a different settlement, may willingly accept a mediator's solution, asking themselves, if not here, then where? When no other solution possesses a prominence that makes it a natural choice as a settlement point, then the endorsement of a mediator may suggest the only settlement point possible (Schelling, 1960; Stevens, 1963; Douglas, 1962).

Multiple-Issue Bargaining Agendas

Many negotiations involve several different issues. Even the traditional negotiation between labor and management involves much more than the size of the final settlement. A typical negotiation between labor and management will involve complex discussions concerning the shape of the final settlement. A union will play a major role in determining how compensation should be split between wages, benefits, retirement, health care, and other fringe benefits. Beyond these monetary issues, workers and management may negotiate over a variety of work-related issues, such as grievance procedures, work rules, staffing, and the rights that workers and management have in determining day-to-day work tasks. In international relations, nations will see single issues, such as arms control, in the context of an ongoing relationship that includes other issues such as trade and actions toward other third-party nations.

This multiplicity of issues is a common feature of bargaining, not just in labor–management and international relations, but in any situation where those meeting share an ongoing relationship. The transformation of a dispute from a single-issue to a multiple-issue bargaining agenda affects both the character of the negotiations and the ability of each side to win partial victories.

How does this multiple-issue bargaining differ from single-issue bargaining? How does the multiplicity of issues affect the bargaining and mediation tasks? In single-issue bargaining, the two negotiating parties need only to determine whether it is possible to find an agreement that improves their lot over the status quo, and then to decide on a settlement. In multiple-issue bargaining, the negotiators may make gains by trading off concessions on one issue in return for gains on another. These reciprocal concessions can allow those bargaining to use differences in preferences to permit both sides to achieve simultaneous gains through bargaining. Pruitt (1981) reports that those negotiations that discuss several issues simultaneously usually facilitate the development of settlements that produce significantly greater joint gains than bargaining that considers each issue in sequence. Thus, the differences in preferences over issues create rewards for those who can discover them.

Suppose that a union and management are negotiating a new contract. Typically, the bargaining will discuss both the total cost and the form of compensation that the final contract will offer. Discussion will often focus on nonmonetary issues, such as work rules and different assessments of the costs of a particular item. Divergence in attitudes toward a particular issue can allow both sides to gain through bargainers. For example, workers may view changes in work rules as a nuisance, while management believes that these changes will enhance worker productivity and management morale. In this case, management might offer wage concessions for the work rules modifications that they value highly, and the union will exchange work rule concessions for the wage increases that they value. This divergence in preferences or assessments allows both management and union to gain by these reciprocal concessions. When the importance of the bargaining issues differs between the negotiating groups, bargaining need not be only a search for an acceptable division of goods or compromise on an issue, but it can allow a cooperative search for settlements that use the different preferences of the negotiators to make everyone better off. Thus, an expansion of a bargaining agenda can offer a way of enhancing the chances for reaching a negotiated solution to a dispute (Raiffa, 1982).

In bargaining over multiple issues, the balance between cooperative and competitive elements changes from that in single-issue bargaining. The negotiators now have three tasks: (1) to cooperatively explore and exploit differences in preferences that allow both parties to improve their lot over the status quo; (2) to advance the interest of their own group; and (3) to determine whether an agreement is possible.

Although a skilled negotiator may use bargaining information both to search for potential solutions and to advance the interest of his or her party, sophisticated bargaining tactics can jeopardize the search for joint bargaining gains. When bargainers are unable to distinguish communications that seek to explore for compromises from probes that primarily seek to determine weaknesses, the entire negotiation enterprise can collapse.

Laws or institutions that enable groups to expand bargaining agendas to include issues over which preferences differ can enable both sides to achieve gains over the potential settlements available in a single-issue dispute. Despite the advantages of multiple-issue bargaining and the inherent complexity of environmental disputes, laws often set single-issue bargaining agendas that accentuate the competitive aspects of a negotiation. In the dispute over the construction of power plants at Colstrip, Montana, the Northern Cheyenne Indians were troubled not only by the impacts of the additional power plants on air quality, but also by the social impacts caused by rapid development by Montana Power. Federal environmental laws seriously constrained the ability of the protagonists to discuss these issues. At the federal level, the dispute centered on the choice of pollution abatement technologies, on the levels of emissions for three pollutants, and on whether to issue a construction permit. The only possible outcomes of this dispute—delay, cancellation, additional pollution controls—would fail to address any of the social impacts of plant development.

The Montana State Facility Siting Act, however, offers an example of a rare law that explicitly calls for a consideration of the impacts of development on local communities and requires planning to control adverse consequences of rapid development. The formal inclusion of these development issues in the public agenda legitimated the eventual quest for measures that would compensate the Northern Cheyenne and enable them to prepare for these "boomtown" impacts. The final settlement, for example, included payments from Montana Power to enable the Cheyenne to expand the size of their police force. This expansion sought to compensate the tribe for increased crime prevention costs and to prepare for the rapid influx of construction workers into the small communities of the Montana prairie.

Many environmental laws create narrow agendas. Disputes become so narrow that they focus not even on the environmental impacts of a particular facility, but on extremely narrow issues, such as whether the plant uses the "best available control technology" or if its "commencement date" exempts it from a particular set of regulations. A rare exception to this pattern is the Hazardous Waste Siting Act of Massachusetts (Bacow and Milkey, 1982). This act seeks to develop state–town cooperation in solving the problem of insuring the existence of an adequate amount of hazardous-waste-disposal capacity. It encourages developers and towns to negotiate a package arrangement including facility-operating and site-monitoring procedures and tax payments that compensate the community for the disadvantages of hosting a hazardous waste facility. Laws such as these can help transform development disputes from a win–lose situation into one that affords both sides an opportunity to make gains while avoiding the stalemate, delay, and uncertainty of current adversarial forums.

As in single-issue bargaining, the mediator can prove particularly helpful to negotiators who realize that many potential settlements would improve their position over the status quo, but who cannot decide on one particular settlement package. In particular, a mediator may help to highlight the common interests that bargainers have in finding a settlement that exploits differences in preferences over the issues (Bartunek et al., 1975; Young, 1970). A mediator may also endorse or suggest a final settlement to both parties. As in single-issue bargaining, the mediator's endorsement can provide the proposed settlement with a prominence that requires acceptance (Stevens, 1963; Schelling, 1960).

In multiple-issue bargaining, a mediator can play a critical role in the exchange of the information necessary to allow the negotiators to exploit differences in preferences to find agreements that leave each better off. The exploitation of relative differences in bargaining preferences requires the exchange of information in an environment of uncertainty and doubt. The communications skills of a mediator can alter the structure of bargaining in substantial ways to facilitate the exchange of necessary information while reducing the risks that a bargaining communication will erode one's position. This topic will be explored more fully in Chapter 6.

Inappropriate Tactics and Commitments

At times, it is not the multiplicity of agreements that limits the ability of the negotiators to find a settlement but rather the tactics of the negotiators. Bargaining actions can actually prevent settlements, even when initial positions and preferences would permit both sides to readily gain from negotiations. When negotiator actions exceed the bounds of civility, then the constructive exchanges needed for bargaining will not occur. Although seldom codified, each bargaining milieu soon develops its own rules concerning the nature of appropriate actions. In labor-management relations, great stock is placed in a bargainer's ability to keep his or her word (Slichter et al., 1975). In international relations, duplicity is more widely accepted but is constrained by implicit codes of conduct (Iklé, 1964).

Unfortunately, negotiations over development disputes have not yet developed such bargaining norms. Although no special norms regulate environmental disputes, U.S. culture has produced general bargaining norms that are readily recognized. In describing inappropriate tactics, Cohen (1980), a student of bargaining, includes (1) extreme initial demands; (2) bargaining with limited authority to make concessions; (3) using emotional tactics; (4) viewing adversary concessions as weaknesses; (5) making stingy concessions; and (6) ignoring deadlines. Cohen labels these bargaining tactics as "Soviet style," suggesting both their inappropriateness and their use by the U.S.S.R. He recommends that one simply discontinue bargaining when confronting these tactics. On the other hand, John Ilich, in *Power Negotiations* (1980), recommends essentially these tactics as a sure way to do well in bargaining. This book, however, offers such a paranoid view of the world that it is difficult to see how one who accepted its insights could choose to live, let alone bargain.

A mediator can serve a particularly important function by supporting the norms of a particular bargaining context (Jackson, 1952; Douglas, 1962). As an outside party, a mediator can psychologically suggest that the wider community observes the actions of the bargainers. Well-known or prestigious mediators can threaten to withdraw from negotiations if bargainers fail to observe norms. Through both implicit and explicit actions, mediators can help induce and enforce appropriate bargaining behavior.

Although limiting tactics to those within traditional bargaining norms can promote settlements, even widely accepted tactics can hinder negotiations. The tactic that can pose the greatest obstacle to negotiations is that of commitment. Through a commitment, the leader of a bargaining group binds his or her organization to a demand that it receive a certain resolution of a particular issue as a price for agreement. In making a commitment, the bargainer limits his or her ability to compromise. Often, this is done by creating a large penalty that the bargainer or group will incur for breaking its commitment. This tactic changes an apparent weakness (limited discretion) into a negotiating strength (Schelling, 1960). When this tactic is successful, the bargainer wins for his or her group the desired resolution of the issue. When, however, a bargainer commits his or her group to demands that cannot be met by the opponent, a bargaining deadlock will ensue. Thus, commitments offer the possibility of a major bargaining victory but carry a risk of creating a bargaining deadlock.

In practice, the risk of miscalculation by a bargainer is greater than a theoretical analysis of bargaining games would suggest. This higher risk stems from the process by which groups make commitments, and from the difficulty of knowing the true nature of the bargaining situation. One of the most common ways of making a group commitment is to tie the group's demands to principles that cannot be compromised. Stevens (1963) notes that this tactic is commonly used in collective bargaining. Those who compromise a principle held by the group incur a loss of respect and authority. In addition, if a negotiator takes unilateral action to modify a position based on principle, his or her group will incur a reduction in bargaining strength, in both current and future negotiations. Bargaining opponents would believe that the negotiator and his or her organization could simply walk away from a bargaining position, and that their bargaining promises are meaningless. The breaking of a commitment based on a principle threatens the credibility of both the bargainer and the organization. Thus, linking a bargaining position to a principle probably creates the strongest commitments.

Another common technique for committing a group is to set the expectations of the group's members for certain outcomes. Once the negotiator-leader changes or sets the expectation of his or her group's members, it may be very difficult to persuade them to accept a smaller settlement package. Committing groups by setting expectations can prove to be a particularly powerful bargaining weapon. In groups where leadership is elected, the failure to meet the expectations of a group can produce the ouster of the leader. The marshaling of membership's opinion can increase a leader's bargaining power by constraining his or her options. When the bargaining opponents can recognize that a negotiator has committed his or her group to a particular set of bargaining demands, they may have little choice other than to concede. If the other bargainers are unable to meet the demands, a settlement may be possible only through incurring the costs of breaking prior bargaining commitments. Unlike commitments backed by principles, however, the costs of disappointing the expectations of a group fall primarily on the leadership.

A sophisticated negotiator may use a combination of these techniques to commit his or her group. However, when bargainers tie an issue to a principle and raise the expectations of their group, they may leave little room for compromise. The 1952 steel negotiations offer an example of how bargainers have used these techniques to commit their groups to positions that can produce costly and prolonged strikes.[2] The 1952 steel negotiations focused on several key issues, including wages, work rules, and union security. During the prebargaining posturing, the industry had taken a principled position against the union demand for a

---

[2]This analysis is based primarily on the account of Livernash (1961).

union shop—requiring workers to join the United Steel Workers of America after they were hired. Industry argued that a union shop was un-American and a violation of a worker's right to freely decide whether or not to join a union. Union leadership argued that management's demands for an open shop and changes in work rules would take working conditions back to the nineteenth century.

This negotiation was complicated by the role of the federal government, which had dominated industrial relations during World War II and the postwar demobilization through a series of wage, price, and production boards. In 1950, after the outbreak of the Korean War, a series of production, credit, wage, and price controls were established. Furthermore, the mobilization effort for the Korean War created a strong national interest in the continuous production of steel. After several months of bargaining between the union and industry, fact finding by the Wage Stabilization Board, and price concessions by the Office of Price Stabilization, negotiations broke down. Fearing the disruption that a strike would have on the war effort, President Truman seized the steel mills on April 8. Steel production continued in government-run plants until June 2, when a Supreme Court decision held that the government seizure of the steel mills was unconstitutional. Within two hours of this decision, a full strike began.

New bargaining began immediately, and by June 8, a tentative settlement was reached on most issues, including wages, but not the union security provision. The steelworkers rejected a management offer to facilitate the union's solicitation of new hires as an inadequate compromise of their commitment to a union shop. Labor and management positions left little room for compromise, and a deadlock was reached. Negotiations halted and did not reconvene until June 20. Still, the union security issue remained the principal barrier to agreement.

On July 23, the Secretary of Defense announced that a shortage of steel was undermining the national defense program. Within a day of this announcement, both sides agreed to a compromise formula: all workers were automatically enrolled in a union, but new workers had the freedom to withdraw during the last 15 days of the first month of employment, and all workers could withdraw from the union in the last 15 days of a contract. This agreement had the practical consequence of placing almost all workers in unions, but it met industry's concern for preserving a worker's right to decide whether to join a union. The compensation package of wages, holidays, and benefits changed little during the strike period. After two more days of bargaining to resolve last-minute issues relating to iron ore workers (who bargained at the same time), the strike ended.

Only the passage of time and the reemergence of national security as a factor in the dispute allowed the negotiators to back away from their commitments. These higher concerns for national interests and fear of public censure superseded the parochial issues tied to competing union–management principles.

In retrospect, it is difficult to say who won the strike or compromised their principles the most. Although industry preserved a worker's right to not join a union, within five years of this agreement union membership levels had climbed from 80% to 90%, and few workers exercised their right to withdraw (Livernash, 1961). All other issues remained virtually unchanged from the June 8 proposals to the July 26 return to work. Thus, a long and bitter strike took place for issues of principle that were only indirectly tied to economic matters.

Both the uncertainty of impacts and the dynamics of forming a commitment can lead negotiators to make simultaneous commitments that eliminate the possibility of a settlement. The process of raising a group's expectations or committing it to a course of action takes time and, once started, may prove hard to reverse or check. In addition, negotiators seldom have complete information concerning the preferences and actions of their bargaining opponent. Each may try to make a commitment to enhance the bargaining situation of his or her group and believe that there remains ample room for agreement. In the steel negotiations of 1952, industry bargainers made a commitment by tying their position opposing a union shop to the principle of worker choice. At the same time, union leadership committed workers to a union shop. In the course of negotiations, the bargainers discovered that their joint commitments had eliminated the possibility of a settlement.

From this example, we can see the interplay of time, principles, and expectations in creating and blurring commitments. Unlike theoretical bargaining games, where players act simultaneously with perfect information, in real life it takes time to make a commitment and it is difficult to predict its impact. Neither side could assess how strongly positions were tied to principles. Neither side (nor government) could predict that such a long strike could ensue from a nonmonetary strike issue.

Unfortunately, in both international relations and development disputes, bargainers often make commitments by tying their positions to principles. Although it is always important to defend deeply felt principles, the frequency of their occurrence in international disputes must raise skepticism over whether they are not simply internal or political stratagems. Similarly, in development conflicts, opposition groups often articulate principles as part of their mobilization strategies, which can leave them committed to positions that preclude bargaining.

Mediator actions or third-party interventions can assist negotiators to gracefully back away from commitments that have eliminated the possibility of a settlement (Pruitt and Johnson, 1970; Stevens, 1963; Schelling, 1960). Even when negotiators realize that commitments have precluded agreement, the commitment places high costs on those that break them. These costs can be both personal, affecting the leader's power and influence within his or her own organization, and institutional, affecting the bargaining power and credibility of the organization itself. A mediator can help to reduce both individual and institutional costs of backing away from a commitment that prevents a settlement. A mediator may suggest an interpretation of a commitment that obfuscates its original meaning (Schelling, 1960). Once the commitment becomes hazy in the minds of the negotiators (or their constituents), only a short time need pass before it dissolves. This pause can allow a negotiator to alter a position so that an agreement can be reached. In this situation, it becomes less certain that the settlement has compromised a principle or failed to meet the expectations of a group's members.

In a dispute over the proposed construction of a dam on the Snoqualmie River in Washington State, a group opposing the original dam tied themselves to the principle of a free-flowing river (Cormick and Patton, 1977). Over the course of negotiations, this commitment appears to have been transformed through reinterpretation. The opposition group's stand was interpreted as a commitment to a free-flowing *Middle Fork* of the Snoqualmie River. The final settlement plan met the flood control needs of the region through the proposed construction of levees along the Middle Fork of the river and the planned construction of a dam on the *North Fork*. The principle of a free-flowing river appears to have been bent, if not outright compromised, through the reinterpretation of the nature and extent of the original commitment. On the other hand, the failure to agree could have produced unilateral action by the governor that failed to address the concerns of the disputants. This obfuscation opened the problem to a consideration of wider range of flood control technologies.

In other situations, a mediator may serve as a scapegoat, enabling the negotiators to blame him or her for forcing them to back away from previous commitments (Simkin, 1971; Stevens, 1963; Kerr, 1954). This procedure can prove particularly important when reinterpretation or blurring of the commitment is not possible. In private, the negotiators may explain to the mediator the difficulties they face because of prior bargaining commitments and the expectations of their groups' members. When necessary, and with the consent of the negotiators, the mediator can publicly pressure the negotiators to move away from their previous positions (Simkin, 1971). Indicting negotiators for being too tough can veil their retreat and guard them from attacks from their own groups' membership (Shapiro, 1970). In the etiquette of bargaining, yielding to an adversary's demands can convey weakness, but yielding to a mediator's request shows statesmanship and reasonableness.

New conditions, such as the announcement by the Secretary of Defense that the steel strike had jeopardized the nation's defense effort in the Korean War allowed both sides to accept a compromise on the union security issue. The negotiators for both sides could, however, avoid the costs of disappointing the expectations of their members by pointing out the necessity of deferring to security interests. Negotiators could return to their constituents and claim that circumstances required the concessions.

Time may prove an important ally to a mediator who seeks to assist negotiators in blurring agreements. Over the course of a strike, new factors may emerge that change the bargaining environment in ways that help produce a settlement. Without a strike, a multiyear contract may permit bargainers to keep commitments in early years yet offer future concessions that make the total agreement more acceptable to their bargaining opponents. A three-year contract may enable management to meet union demands for the first years of a contract, yet meet management commitments through savings in the later years (Simkin, 1971). Union leadership can stress to their membership the benefits they gain immediately, while management can stress the overall cost of the contract. This technique, however, is seldom available when a principle is at stake.

Discrete Bargaining Issues

Not all the issues in negotiation will offer the bargainers continuous choices. In labor–management disputes, although wages can vary almost continuously, work rules, which govern the production process and the relation of labor to management, often appear to permit only discrete choices. Firms either have a union shop or they do not; technologies are either used or they are not. Negotiations over such discrete issues as union security, the introduction of a new technology, or new machine-staffing levels can produce bitter strikes that will end only when one side concedes on the resolution of the discrete issue (Simkin, 1971).

International conflicts are often affected by the discrete character of major issues. Territory usually creates a discrete issue—either the territory is occupied or it isn't. Even when disputed terrain appears to permit division, rivers, mountains, oceans, and deserts often give territory a discreteness which featureless maps do not depict (Schelling, 1960). These issues of territorial possession are among the least likely to be resolved through negotiation or compromise.

Unfortunately, many of the issues of an environmental-development conflict have a discrete or polar character. Rivers are either "free-flowing" or they are not; an area is either a "wilderness" or it is not. Although sometimes the discrete character of an issue will facilitate an agreement, more often this discreteness will present a difficult negotiation barrier. The art of negotiation generally requires that there be no clear winners or losers. It works best when both sides win. If an issue does not permit compromise or obfuscation, then only one side's concession or capitulation to the other will resolve the dispute. The resolution of discrete issues will often necessitate the creation of clear winners and losers.

A common technique for resolving disputes about a discrete issue is to seek to link that issue to others that permit one bargainer to compensate the other for a major concession on the discrete issue. Through reciprocal concessions on linked issues, both sides can obtain bargaining victories that permit them to reach agreements while avoiding destructive conflicts. When negotiators do not link issues themselves, a mediator may facilitate linkages between a discrete issue and other issues at the bargaining table by exerting control of bargaining communications and the negotiation agenda. Setting a negotiation agenda with several issues on it will prove more conducive to linking issues than setting an agenda that addresses each issue in sequence. Mediators can help unskilled negotiators to set a bargaining agenda with issues that facilitate reciprocal concessions.

When linkages do not offer a productive way of negotiating, mediators may help the negotiators to create alternatives to the polar or discrete issues, often by casting them into an environment that permits a wider choice of alternatives. Conflicts over discrete issues often arise because of the institutional arrangements between the negotiating parties or because of the technology for a production process. In labor disputes, mediators sometimes have a wider range of experience in negotiations and labor–management relations than the parties bargaining. This experience can enable them to suggest new ways of resolving issues that appear polar to negotiators. Simkin (1971) describes how the mediator of a West Coast dock strike realized that a union was fighting the introduction of a new technology because it feared the loss of jobs, not out of any aversion to progress. The union insisted on retaining work rules that would limit the use of this new technology and would require high staffing levels. Management believed that it had the right to introduce work techniques that increased productivity and lowered costs. The mediator placed this problem in the broader context of job security and productivity. He successfully suggested a compromise solution in which workers received job security and wage increases in exchange for cooperation with management in the introduction of new technologies.

Mediators may also recast a discrete issue into a continuous form. In a dispute over the designation of a wilderness area in Idaho, the mediator, with the assent of the conflicting parties, transformed the conflict from one that centered on whether or not to have a wilderness region, into a land-use planning exercise that decided on the size of the wilderness region. The conflicting sides worked together to set aside some regions as wilderness, while others were freed for logging (*Congressional Record*, 1977).

Mediators who can envision new relationships between developer and environmental groups will be particularly important in resolving conflicts over the construction of new facilities. Through the negotiation process, a mediator can try to create a climate of cooperative problem-solving in which the parties discuss interests and goals, rather than positions. The mediator's success may depend on an ability to see compromise solutions and to create new dimensions in issues that appear polar to the disputing groups. In development disputes, a mediator will profit from a knowledge of the alternative ways of designing a development project. In the dispute over a proposed dam on the Snoqualmie River, the participation of an engineer with a knowledge of the technology of flood control enabled the negotiators to devise an alternative to the Middle Fork dam. The suggestion of constructing a levee on the Middle Fork (in combination with a dam on the North Fork) to provide flood control offered a feasible and needed alternative (Cormick and McCarthy, 1974) that could provide flood protection without eliminating the free-flowing river.

Of course, not all conflicts involving discrete issues can be resolved through linkages with other issues or combinations of other issues. Negotiators may find that when a single polar issue dominates a conflict, bargaining cannot produce a settlement. When there exists no compromise sufficient to compensate a bargainer for a concession on a discrete issue, then the search for a settlement can proceed in several ways. The negotiators can search for still more issues to link to the discrete issue, hoping that some compromise on several issues will compensate for a concession on the discrete issue. If this is not possible, negotiators can try to develop some new alternative resolution to the discrete issue that occupies a position intermediate to the original extremes. Furthermore, events, mediator action, or new information may alter the preferences of the negotiators over the disputed issues so that a settlement becomes possible. Nevertheless, when one single discrete issue dominates a dispute, a settlement may depend more on luck than on the skill of the negotiators and may end more often in disagreement than in agreement.

Conclusion

The number and nature of the issues in a bilateral bargaining situation alter both the negotiation and the mediation tasks. Single-issue bargaining requires negotiators to make concessions in a competitive environment in which one side gains what the other loses. In negotiations over several issues, differences in preferences over the issues between the bargaining organizations can permit both negotiators to realize gains over the status quo. Thus, bargaining includes both a competitive and a cooperative element. A cooperative search is needed to discover those resolutions that can improve both sides over the status quo, but each bargaining group competes with the other over the choice of any one particular settlement:

- Those who hope to promote a negotiated resolution of development disputes should try to establish a negotiating agenda sufficiently rich to permit a cooperative search for improved agreements.

Even when bargainers can successfully find many acceptable agreements, they may fail to agree on one. In those cases, a mediator may prove particularly helpful:

- When the multiplicity of potential agreements limits the ability of the negotiators to reach a settlement, a mediator can help the negotiators by endorsing a compromise solution. The mediator's endorsement will confer a prominence on one particular settlement that will lead to its adoption.

The tactics used by the negotiators can affect the likelihood of reaching a settlement. Tactical commitments, whereby a group attempts to gain a bargaining advantage by tying itself to a particular position, carry the risk of eliminating the possibility of achieving a bargaining settlement. In negotiations fraught with uncertainty and poor information, bargainers may eliminate the possibility of settlement through simultaneously committing their groups to demands or positions that leave no room for agreement.

Table 4
Tactical Aspects of Bargaining Systems

	U.S. Industrial Relations	International Relations	Development Disputes
Size of agenda	Multiple-issue.	Often one issue.	Often one issue.
Norms for tactical behavior	Lying limited by codes of behavior. Norms of conduct developed as part of an ongoing relation.	Diplomatic norms provide little guidance on bargaining tactics.	Norms not set, but societal bargaining norms are often shared.
Commitments	Often arise from the actions of bargaining leaders. Usually internal politics leave leaders free to act.	Can often develop from internal politics.	Often arise because the public position of the group and its membership are tied to principles.
Nature of bargaining issues	Wage issues are continuously divisible. Issues over work rules often have a discrete character.	Major issues are indivisible.	Major issues are indivisible.

- When commitments eliminate the possibility of a settlement, a mediator may play a key role in allowing the negotiators to back away from commitments that prevent them from reaching a settlement. A mediator can allow negotiators to gracefully back away from commitments by offering the negotiators a "statesmanlike" compromise or by offering himself or herself as a "scapegoat" who forced a bargainer to concede.
- When commitments preclude settlement, a mediator may help negotiators to obfuscate their commitments by providing new interpretations of previous bargaining positions or by offering new information that alters the bargaining environment.

The structure of the issues in a dispute can affect the likelihood that bargainers can resolve them through negotiations. In particular, when negotiators are locked in a dispute over a discrete issue, the chances of a negotiated settlement can decrease. Often, a settlement becomes possible when the negotiators can link discrete issues to those issues that permit the bargainers to make reciprocal concessions. In addition, negotiations and discussion may permit the bargainers to envision new institutional arrangements or resolutions that compromise issues that were initially conceived to possess a polar form. Although mediators need not take the lead role in helping bargainers to overcome the obstacles posed by discrete issues, they can help in several ways:

- When the discrete character of an issue complicates the search for a settlement, mediators can suggest linking issues so that one side can compensate the other for concessions on the discrete issue.
- When the linkage of the discrete issue to other issues does not permit the negotiators to resolve their differences, the mediator may help the negotiators to design some new alternative that offers a compromise over the original polar form of the issue.

Table 4 summarizes our analysis of how the bargaining agenda, the norms for tactical behavior, the source of bargaining commitments, and the nature of bargaining issues vary over these three contexts. Once again, one can see how laws and behavioral norms support the collective-bargaining process. Unfortunately, neither international relations nor the current character of development disputes possesses those elements that clearly aid industrial relations to resolve a dispute. In these contexts, mediators often provide the assistance that the bargaining environment does not. For disputes over development projects, changes in law can create structures that promote negotiations.

CHAPTER 6

Information and Communication, Negotiation, and Mediation

Introduction

It is through communication that bargainers reach settlements. Effective communication provides several positive benefits to bargainers. Communication can enable negotiators to exploit the cooperative elements present in bargaining. A communication of basic interests can permit bargainers to develop new alternatives that offer both sides gains. When several issues are present, communication can permit bargainers to exploit differences in preferences in designing agreements that beat other alternatives.

Bargaining contains many competitive elements, and communication can prove essential to discovering and exploiting an opponent's weaknesses. Negotiators often possess little information and harbor doubts concerning their own power and their opponent's strengths and interests. Information is not only scarce, but often possessed asymmetrically. These asymmetries can introduce competition, fear, and doubt into bargaining. This chapter considers:

- How partial information alters the incentives to bargain in single- and multiple-issue negotiations;
- How asymmetries of information can inhibit the exchange of the information necessary for bargaining;
- How mediators can limit the vulnerability that bargainers face in making truthful bargaining communications.

In development disputes, asymmetries of information occur constantly. The developer will always possess the best understanding of the cost, design, and nature of project alternatives but will have a less concrete understanding of the effects of development on a community. Community groups, on the other hand, often lack the expertise needed to evaluate available information and the resources to develop their own but have a detailed understanding of the impacts on a community. Thus, information will always play a particularly important role in these negotiations.

Information Generates Bargaining Power – Single-Issue Bargaining

Consider the following story: In many movies set in the Far East, pearl buyers wear dark glasses. This habit is not merely dramatic convention, nor is it the aberrant behavior of shady characters. The tropical sun is bright, but the merchants wear glasses to protect their eyes not from the sun's rays, but from the probing gaze of others. When a buyer sees a pearl of uncommon beauty, the pupils of his eyes dilate reflexively. The seller who sees this knows that he can get a high price. The buyer wears the dark glasses to conceal this information.[1]

This story of pearl trading describes a simple situation of bargaining over a single-issue, the price of exchange, but it illustrates many of the major features of the information structure that commonly surrounds bargaining. In this story, a buyer and seller wish to strike a bargain over the pearl. The seller will have a minimum price below which he will refuse to sell the pearl; the buyer will have a maximum price above which he will refuse to buy. As mentioned in the last chapter, these prices are called *reservation levels*. When both are well known (as with the *Kelley Blue Book* auto prices), splitting the difference proves natural. In this real-world example, neither bargainer knows the reservation level of the other. Through bargaining, they must determine whether a bargain is possible and, if possible, then agree on a price.

If the seller's and buyer's reservation prices permit a sale of the pearl, then asymmetrical information about reservation levels has great strategic value. If the seller knows the highest price that the buyer will pay while the buyer does not know the seller's minimum price, they should be able to strike a bargain at a high price. Similarly, the buyer

[1] I remember reading a version of this story years ago, but I cannot recall the source.

who knows the seller's lowest acceptable price should do very well. Both participants will try to get information, either from outside investigations or from the bargaining communications. If the seller stole the pearl (as these old movies often imply), he may have a poor idea of its value. The theft, however, prevents him from getting the pearl appraised by a legitimate merchant. When a bargainer does not know the minimum price that his opponent will accept in exchange for the pearl, he will likely make an estimate of what value it takes. Any information will help the bargainer estimate the true value of the pearl. Of particular value is information that the bargainer knows is true.

The buyer must see the pearl to know what to offer. In bargaining over the pearl, the reflexive dilation of the buyer's pupils gives true information concerning his evaluation of the pearl's worth. Note that the buyer's reaction to a beautiful pearl does not reveal his highest offer, but it gives information that the seller may use to revise his estimate of the buyer's highest offer. A streetwise seller will raise his price when he sees the buyer's eyes dilate. The buyer wears the dark glasses to prevent the seller from gaining this information and its potential bargaining advantage. By wearing sunglasses, the buyer can break the link between his examination of the pearl and the information transmitted by the reflex of his eyes.

In any negotiation, the bargaining communications between the buyer and seller provide a source of information concerning reservation levels. The pattern of the demands and concessions of bargaining, and even the facial and body expressions of the negotiators, may convey information of strategic value. The level of the opening demand, the size of concessions, and the time between offers can reveal to an experienced bargainer the desire of the opponent for an agreement (Pruitt, 1981).

Unlike the reflexive dilation of the eyes of the pearl buyer, bargainers can control much of this information and may attempt to use their bargaining communications to manipulate the expectations of their opponents. The expectations of bargainers depends in part on what they believe they can attain (Kelley, 1966). When bargaining skills are asymmetrical, the better or more experienced negotiator may manipulate these communications more effectively and win a settlement more favorable to his or her interests. The less adept bargainer may be unable even to suppress his or her reactions to the statements of the bargaining opponent. When bargainers have comparable skills and experience, the bargaining outcome will more likely approach the 50–50 division suggested in our earlier analyses of the strategic structures of bargaining.

Information Generates Bargaining Power — Multiple-Issue Bargaining

Bargainers often negotiate over not one issue, but several. When bargaining, negotiators neither have complete knowledge of their opponent's preferences nor know for certain the actions their opponents have taken, or the truth of what they say.

The fact that bargainers have only partial information regarding preferences and reservation levels introduces cooperative elements into multiple-issue bargaining. As in single-issue bargaining with partial information, the bargainers have an incentive to cooperate to determine if any agreement is possible. Unlike in single-issue bargaining, even if two negotiators find a resolution of issues which they both can accept, there may exist other settlements which both would prefer to the one that they have selected. Each can try to discover whether preferences permit them to find some new arrangement of the issues that could leave him or her better off. This opportunity creates an incentive for the bargainers to cooperate with each other and to reveal how they are willing to trade concessions on one issue for gains on the other. The negotiation search will have reached an efficient settlement only when no combination of reciprocal concessions exists which can leave them both better off.

Partial information also creates incentives to compete. Just as information about the reservation price gives the pearl buyer a bargaining advantage, the possession of information in multiple-issue bargaining confers a strategic advantage. In bargaining, a negotiator will try to estimate the preference structure of the bargaining opponent, and to determine which settlement will most advance his or her own interests (Raiffa, 1982). Furthermore, a bargainer who is unsure of the true value of his or her good or service may use negotiation communications to manipulate the preferences and demands of the opponent. Thus, the bargainers will have competitive incentives to try to determine or shape the preference structures and reservation levels of their opponents and to use this information to advance their own interests.

An awareness of the competitive aspects of bargaining and of the strategic value of information creates an incentive for a bargainer to attempt to mislead the opponent. A reflective bargainer, in examining the bargaining situation from the viewpoint of the opponent, will realize that the opponent also has an incentive to try to determine his or her preference structure and reservation levels. After gaining this insight, this bargainer not only may use negotiations to learn information concerning the opponent but also may attempt to manipulate the opponent's perceptions of his or her own preferences and reservations. Providing the opponent with either selective or misleading information can help the bargainer to create perceptions that will give him or her a bargaining advantage in negotiating the settlement.

When bargainers adopt a cooperative approach to negotiations, one common procedure is to match the bargaining concessions of an opponent (Pruitt, 1981). Without information concerning the opponent, it is often difficult to judge the cost or value that an opponent places on a particular concession. This lack can allow a skillful bargainer to create a situation in which minor concessions elicit major rewards. The bargainer may seek to convince the opponent either that his or her reservation level for a settlement is higher than it actually is, or that his or her preferences over the two issues require major concessions by the opponent for concessions that are minor. In multiple-issue collective bargaining, for example, each side would like to trade concessions from its "wish list" for gains on the core issues. The inclusion of these nonessential issues may actually help negotiators to improve their final settlement by facilitating linkages. A negotiator will, however, attempt to sell concessions for a high bargaining price and buy concessions for a low price.

When both bargainers analyze the negotiation situation fully, each will realize that the opponent may try to manipulate his or her perceptions. The levels of strategic interaction can quickly mount so that communication carries little information. Each bargainer will discount highly the statements and information provided by the others. However, when the negotiators differ in their bargaining skills, the more skillful negotiator may have an advantage in managing bargaining communications and the exchange of information.

All incentives to mislead are tempered by a negotiator's fear of losing a potentially beneficial agreement. If the parties have little information concerning their opponent, bargaining demands may be mismatched (Pruitt, 1981). Low initial demands or concessions may be interpreted as weakness and elicit high counterdemands that the bargainer cannot meet. When an agreement is not necessary, high initial demands may just drive the other bargainer away. For example, asking too much for a new or used car may just drive the buyer to another lot. Furthermore, the search for an agreement requires time, energy, and resources. A great divergence in bargaining positions may simply cause the negotiators to decide that the likelihood of a beneficial agreement is so small that it fails to justify the commitments of time and resources which negotiation requires. Mismatched bargaining demands heighten the competitive dimensions of bargaining. If the divergence in positions stems from a bargaining strategy rather than from the underlying preferences, the negotiators may lose beneficial agreements.

The tactical nature of commitments also changes under partial information. Since the preferences and reservation levels of a bargaining opponent are not known, a negotiator may make a commitment to improve his or her bargaining position and inadvertently eliminate the possibility of agreement. Further, the lack of information may leave bargainers unaware that their commitments have destroyed the possibility of an agreement until they have expended a great amount of time and effort searching for one. A bargainer may also have a difficult time communicating his or her commitment to the opponent. In practice, it may be difficult for a negotiator to distinguish between a nonnegotiable commitment and a negotiable bargaining demand. Additionally, for strategic reasons, a bargainer may attempt to interpret commitments as negotiable demands. This misinterpretation may obfuscate the commitment and convert it into a negotiable demand. In these ways, the lack of clear commitments may complicate the bargaining structure.

Bargaining Communications, Information, and Negotiation Failures

Bargaining communications can provide a negotiator with information concerning the bargaining opponent. The strategic value of information revealed in bargaining communications depends on the amount of new information revealed and the relative skills of the bargainers (Raiffa, 1968).

In many bargaining situations, the sides know each other well. They may have other sources of information outside the negotiation framework which will give them a fairly clear idea of the reservation levels and bargaining preferences of their opponents. This information reduces the strategic value of the information which the negotiators may reveal in the bargaining. For example, when the United Auto Workers and General Motors negotiate, each has a great deal of information about the other. In particular, there exists a history of settlements between the two sides. Further, labor will know the size of the industry profits over the last year and the size of current inventories. Management will know the size of the union's strike fund. Both sides will have outside information concerning the rate of inflation and wage settlements in other industries. Thus, existing information may help narrow the range of key issues and help each side form realistic bargaining expectations. The strategic value of bargaining communications is likely to be small.

In situations in which each side knows relatively little about the reservation levels or preference structures of the other negotiators, the bargaining communications may carry much information of strategic value (Harsanyi, 1972). In bargaining, the negotiators may seek to communicate their preferences over the issues and explore the possibilities for new agreements which offer gains. When bargainers communicate their preferences, they also give the other negotiator some information concerning their entire preference structure. Just as noticing the slope of the land as one hikes suggests whether one is walking in a valley or on a ridge, information concerning local trade-offs between issues helps an opponent determine the shape and boundaries of one's preference. Such knowledge can enable a skilled opponent to gain accurate knowledge of one's minimal demands. Even when a used-car dealer asks you, "What type of car do you have in mind?", your answer will convey some information concerning how much you are willing to spend.

The bargainers face a problem. To bargain effectively, the bargainers must determine whether a settlement is possible, and whether any other potential agreements would allow them to make each other better off without making either worse off. On the other hand, each side must guard against conveying information about its reservation levels and preferences that the other side can exploit to win a settlement most favorable to its interests. Unfortunately, those messages that convey information about preferences over potential concessions almost always convey some information concerning reservation levels and bargaining expectations. Thus, a bargainer needs to find some mechanism that signals to the other bargainer productive areas to explore for possible agreements without revealing information that may jeopardize his or her ability to win a favorable settlement.

Pruitt (1971) analyzes the effects of communications on bargaining in psychological terms that closely correspond to the strategic ideas of reservation levels and bargaining preferences. He makes the distinction between position loss and image loss. Position loss stems from the fact that once a concession is made or even suggested, it is impossible to completely withdraw it, and its bargaining value is reduced. Thus, it can be very helpful in negotiation to be able to signal to a bargaining opponent possible bargaining concessions without explicitly offering them.

Bargaining image is the way that an individual is perceived by the participants in bargaining (Pruitt, 1971). An image loss can arise from a bargaining concession if it suggests that a bargainer is more flexible than he or she at first seemed, or that his or her bargaining position is weak. The perception of weakness can lead a bargaining opponent to press his or her demands more strongly, or to attempt to gain a bigger bargaining victory.

Fear of position loss or image loss can prevent needed communications. Pruitt (1971) sees the solution as establishing a mechanism for making simultaneous bargaining concessions, arranging for sequential exchanges of concessions, or "sequential inching" toward a settlement, whereby very small concessions are exchanged in a long negotiated dance toward agreement.

Pruitt (1971) argues that at times these direct approaches to bargaining may result in premature position loss. Revealing one's bargaining hand early in the negotiations can permit the opponent to exploit the strategic content of the bargaining messages. Pruitt suggests that forms of indirect communication can allow a negotiator to signal that he or she may make a concession if he or she has assurance that the bargaining opponent will make a comparable one. This indirect communication allows the bargainer to avoid both position loss and image loss. Pruitt identifies four types of indirect communications: (1) tacit communication; (2) informal conferences between the negotiators; (3) messages transmitted through intermediaries; and (4) communications through mediators. These four mechanisms offer a way of exploring the bargaining terrain without yielding information of strategic value.

Tacit Communication and Informal Conferences

Tacit communication includes a variety of verbal and personal clues with ambiguous meanings, known best to the different bargainers. These signs get their meaning from the familiarity of bargainers with each other and from the ability of the bargaining opponents to interpret these signs (Douglas, 1962). By sending an ambiguous message, one does not commit one's side to a particular position. Pruitt (1971) states that a norm of truth telling pervades tacit communication. Although the communications are intentionally ambiguous, the parties neither lie nor overstate their positions in the tacit communication channel as they may do in formal negotiation sessions.

Once again, labor–management relations offer many examples of how tacit communications can become incorporated into the overall bargaining relationship. Ann Douglas, in *Industrial Peacemaking* (1962), analyzes collective bargaining in four different cases. Her analysis reports that collective bargaining passes through three stages, in which much of the relevant action lies not in what is said at the bargaining table, but in what is left unsaid. The bargainers first define the range of bargaining issues, then attempt to determine the issues available for compromise, and finally, conclude with actual decisions and either agreement or strike. During the first phase, each side establishes its own demands and defends them against the attacks of the other side. This blustering and rigid presentation determines the range of bargaining issues. Introduction of additional issues after this stage is considered inappropriate and grounds for forcing a strike. During the second phase, discussion shifts as each side begins to discuss the elements in the other side's bargaining proposals rather than its own. This discussion can signal the issues that will provide the potential reciprocal concessions for the final stage of bargaining. The final stage produces the actual decision. Douglas observes that although neither side says directly what it means in the early stages of bargaining, it is clear that the exchanges follow rules that are well understood by the negotiators and that each side has normal expectations for a certain progress. The negotiations, though tacit, fulfill the crucial need of setting a negotiation range and signaling where concessions can occur.

In my view, tacit communication is a parallel communications channel that coexists with the formal communications that go on across a bargaining table. Pruitt's examples are drawn from negotiations between parties that have a long-term standing relationship. In any negotiation, there is a need to both cooperate and compete. Any negotiation over multiple issues needs a mechanism for channeling the search for agreement into an exploration of productive issues. The tacit communication channel has arisen to fulfill these cooperative needs. Such a bargaining channel can greatly improve the efficiency of a search for a bargained settlement. As pointed out before, a bargaining competitor can easily exploit such communications. Hence, this important and efficient channel arises with a norm of truth telling. Such a norm is enforced through the ongoing relations between the negotiating parties and through the premium put on reputation in bargaining, which is especially true in collective bargaining. I think it may be most productive to consider the existence of tacit bargaining as a convention that arises within an ongoing bargaining relationship that follows a set of rules enforced by the participants. The bargaining messages are not indeciferable to the participants. If so, they would have little use and a norm of truth telling would be unenforceable. Rather, they are important aids in the search for a cooperative solution.

In addition to tacit communications, Pruitt (1971) suggests that informal conferences can help bargainers to explore for potential settlements. Iklé (1964) states that in international negotiations, informal conferences take place with two norms: (1) what takes place at the conference is secret, and (2) concessions made in informal conferences can be withdrawn. In labor negotiations, Douglas (1962) also records how a labor mediator was able to use the technique of formal and informal conferences to help negotiators in a collective-bargaining dispute reach an agreement. The mediator would explore the potential for settlement in an informal meeting that included both parties. If a potential compromise developed, the mediator would then hold a formal bargaining meeting to get the settlement on the record. Douglas noted with some surprise that the negotiators would go over in a formal meeting the same ground that they had explored in informal meetings while acting as if the informal sessions had never taken place. Pruitt (1971) also states that norms of speaking truthfully, a willingness to reciprocate concessions, and honoring informal agreements in formal meetings that mark tacit communication also hold in informal conferences.

The ability of negotiators to use informal conferences to search for a possible settlement depends on both trust and skill. Negotiators must trust that the other bargaining side will not use the communications given in these informal sessions to exploit bargaining weaknesses that these communications expose. In addition, negotiators must possess the skill to distinguish between formal and informal sessions, and to understand how the informal sessions supplement the formal. Past experience with the use of both formal and informal sessions can greatly aid the negotiators to understand when they are going off the record, and when the session is a formal one. Pruitt (1971) notes that diplomats must learn what different modes of expression indicate about the nature of the bargaining session. As in tacit communications, informal conferences establish a different channel of communications that can enable parties who trust each other to explore the bargaining range to determine whether there exist potential arrangements that can improve both sides.

These mechanisms for establishing communications are the product of ongoing and mature bargaining relations. In collective bargaining, individual bargainers may face each other across the bargaining table many times over the years. This familiarity creates a climate in which informal communications can arise. Repeated meetings help enforce truth telling and enable negotiators to acquire the skill to decode tacit signals. Furthermore, information about the strategic position of either side is likely to have limited value since opposing groups with ongoing relations will have substantial information concerning each other. In international negotiations, formal diplomacy provides support and facilitates the establishment of informal channels. Institutions such as the UN allow the diplomats to have contacts with diplomats from countries even when no formal relations exist.

When negotiators have little information concerning each other's bargaining situation, they may especially fear that their opponents will gain great advantages by exploiting bargaining communications. This will be particularly true in development disputes, in which individuals are usually bargaining on a one-time basis. When opposing negotiators adopt a style that relies on threats and intimidation, an individual may be especially reluctant to engage in bargaining. Furthermore, new bargainers will lack the history of a bargaining relationship that makes tacit bargaining possible. They may not trust that bargaining opponents will honor the norms of informal conferences or may not possess the skill to make the distinctions between informal and formal conferences. In such situations, even when agreements are possible, bargainers may still need a mechanism that allows them to signal their preferences among the issues without tipping off the opponent to their bargaining strengths and weaknesses, to their preferences and reservation levels, or to their fears of loss of position or bargaining reputation. The next section explores how a mediator, like the sunglasses of the pearl buyer, can suppress strategic information yet facilitate essential bargaining communications.

Mediators and Bargaining Communications

Once a mediator enters a negotiation, he or she creates an additional channel of communication between the bargainers. In addition to talking directly to the bargaining opponent, a negotiator can send messages through the mediator. This communication channel offers both new and experienced bargainers several advantages over face-to-face negotiations. For inexperienced bargainers, the physical distance from the opponent that working through the mediator affords the bargainers can help them to hide their inexperience and fears (Pruitt, 1971). Furthermore, it can screen novice negotiators from tactics aimed at intimidation and may perhaps save the negotiations from destructive consequences that unchecked emotional reactions to the stresses of bargaining can generate.

A mediator may also help even experienced bargainers to overcome asymmetries of information and the consequences of bargaining messages. The participation of a mediator can help negotiators to explore settlement possibilities which, if directly proposed by a bargainer, would erode the bargainer's position. This message channel can be of particular importance when a negotiator wishes to explore settlement possibilities. Negotiators may feel that if they make an offer, the very act of making it will signify that they do not value this issue very highly or may cause their opponent to consider their concession a sign of weakness (Stevens,

1963). Since a mediator can explore the acceptability of a particular agreement with the bargainers without signifying whether it was his or her idea or the other bargainer's, negotiators can explore potential settlements without incurring the costs of offering a concession (Pruitt, 1971).

Perhaps most importantly for experienced bargainers, a mediator can facilitate the search for reciprocal concessions and permit the negotiators to make contingent bargaining offers and simultaneous bargaining moves. Simkin (1971) reports that it is extremely difficult for a bargainer to explore a contingent offer without making a commitment to the concession. Simkin then describes a process in which a mediator, through confidential discussions, gains concessions from both sides. The mediator undertakes the search for new solutions as part of routine issue explorations in which he or she suggests many potential solutions. The search for a specific bargaining move may either begin with a side's confidential offer to the mediator of a contingent concession or with a mediator-initiated suggestion. Once the mediator has obtained the approval of both sides to a new bargaining position, the agreement may be formalized in a joint bargaining session. Douglas (1962) offers numerous examples of mediators arranging for simultaneous concessions over the course of negotiations. In this way the mediator directly uses his or her communications functions to facilitate a bargaining agreement by breaking the link between strategic information and information critical to the negotiations.

When a mediator explores the possibility of a particular reciprocal concession with a negotiator in a private session, the bargainer can never be sure whether the offer of a possible concession originated with the mediator or with the opposing bargainer (Jervis, 1970). The uncertainty concerning the source of the offer weakens the link between information concerning a bargainer's preferences on issue trade-offs and information which would assist a bargaining opponent to estimate the other's reservation levels. In negotiations, the mediator acts like the sunglasses, which, in our pearl exchange, suppress strategic information. However, unlike the sunglasses, the mediator's actions may help both bargainers, rather than just one. Because of his or her ability to screen out strategic information, both sides may give a mediator information in confidence which they would otherwise withhold from bargaining (Pruitt, 1971).

For the mediator to work effectively as a screen, each side must trust that the mediator will honor the confidentiality of communications. In current labor–management relations, institutional structures support the confidentiality of communications and the neutrality of the mediator. The Federal Mediation and Conciliation Service, as established by the Taft-Hartley Act, creates an incentive structure for the mediators that is very different from those that the negotiators face. The mediator has an incentive to help both sides to reach an agreement, but no incentive to act on behalf of one particular negotiator or a particular cause. The mediator's salary is paid by the federal government, and he or she owes nothing to either side. Professional ethics produce incentives for a mediator to respect confidential communications.[2] Negotiators can expect that a mediator will use the information they give to him or her about their desire for an agreement to shape the final terms of the settlement, but they do not expect that the mediator will exploit these revelations as a bargaining opponent would. Additionally, a concern for his or her professional reputation among the corps of mediators may prevent a mediator from using unethical means to reach a settlement.

Managing Single-Negotiation-Text Bargaining

One technique that mediators can use to facilitate the search for a settlement is single-negotiation-text bargaining (Fisher and Ury, 1981). In bargaining from a single negotiating text or the status quo, the negotiators start discussion from a single text (rather than from their initial positions) and search for modifications of the initial text that provide each side with gains. Through his or her confidential communications, a mediator may learn the preferences of each negotiator over proposed changes in the tentative agreement. Using this information, which can take the form of criticisms of a proposed text, a mediator may suggest hypothetical changes to the bargainers individually and confidentially. When he or she gains the assent of both to a tentative modification of one part of the text, the mediator can announce this and begin the process again with another issue or bargaining point. This process continues until the negotiators cannot find any more improvements. The negotiators then decide whether they are willing to accept the results as a settlement. Through this procedure, the mediator may assist the negotiators to move toward an agreement through a series of small bargaining steps that provide joint improvements over any tentative agreement.

A single-negotiating-text bargaining managed by a mediator can overcome many of the barriers posed by the information and communications structure of negotiations. Fisher and Ury (1981) suggest that this technique can often succeed because it does not initially ask either side

[2]See, for example, the "Code of Professional Conduct for Labor Mediators," adopted by the Federal Mediation and Conciliation Service and the several state agencies represented by the Association of Labor Mediation Agencies, reprinted in Simkin (1964).

to give up a position, and because it is always much easier to criticize a text than to offer a new negotiating position.[3] Thus, this technique does not require negotiators to incur the high costs often involved in getting a complex organization to agree to a new bargaining position.

From a strategic view, this procedure may limit the information concerning a negotiator's preferences and reservation levels that is revealed through bargaining communications. In bargaining from a single negotiating text, a mediator approaches each negotiator in confidence to gain his or her consent to a bargaining step. The mediator announces to both negotiators when they have accepted a bargaining step. Since each bargaining step is small, the tentative consent of the negotiators to a new negotiating point may reveal information in quantities small enough to limit the possibility of future strategic exploitation. Furthermore, the negotiators know only when both have agreed to a bargaining step. They receive no information which tells them who suggested the bargaining step, and they do not learn which bargaining suggestions were posed by a mediator but rejected by the other negotiator. This single-negotiation-text technique follows the "in the pocket" mediation tactic described by Simkin (1971), through which a mediator gains a concession from one side contingent upon the other side's concession, but puts this offer in his or her pocket until he or she later wins the required concession. By shielding the source of the bargaining offers until they have been accepted, this technique greatly limits the amount of information available to the negotiators.

This mediation or bargaining technique makes it unnecessary for a bargainer to reveal what constitutes an acceptable agreement until the bargaining is over. Bargaining proceeds until the negotiators fail to discover any steps that will allow them to realize joint gains or until they reach a deadline. At that point, the bargainers can decide whether or not to accept the proposal on the bargaining table as a settlement. Only at the conclusion of bargaining do negotiators learn whether an agreement meets their opponent's reservation levels.

This technique is aided by traditional bargaining norms and ethics. Bargainers commonly accept the tentative steps toward settlement as irreversible (Douglas, 1962; Simkin, 1971; Stevens, 1963). Once both accept a bargaining step, neither will renege. The irreversibility of the bargaining steps acts to limit the possibility of strategic action based on information revealed through bargaining communications. Without this taboo, bad-faith bargainers could negotiate once to explore the other

---

[3]Raiffa (1982) describes how President Carter used this technique at Camp David to mediate between Sadat and Begin.

side's preferences and weaknesses and then exploit them in subsequent negotiation sessions. In her case studies of collective bargaining, Douglas (1962) records how collective bargaining broke down when one side introduced issues at the second step of negotiations that were not raised in the initial step, which, by tradition, sets the bargaining range. The irreversible process that facilitates the revelation of new information does so only as the mutual acceptance of the text (and bargaining step) limits the strategic value of the information. This process can limit the ability of a skilled negotiator to use bargaining information to exploit an inexperienced bargaining opponent, yet it is the very procedure that a group of skilled negotiators would insist be followed.

To bargain from a single negotiating text, both sides must express a willingness to try out the process and accept a tentative starting point. The choice of the starting point, however, can influence the final outcome (Raiffa, 1982). It will set the negotiation issues and serve as an anchor to discussions. As a result, the formulation of a starting text may pose a difficult problem for a mediator.

In some bargaining situations, there may be a natural starting point which both sides will accept as the negotiating text. Often, the status quo will provide such a starting point for negotiations. When the setting of the conflict provides a natural starting point, the mediator need not generate it. Cormick and Patton (1977) describe a dispute over the number of lanes of a highway in Washington State. In that dispute, the existence of a road along the proposed highway route served as a status quo around which all parties sought gains. A five-lane road existed which many considered unsafe. Although some opponents of the proposed development project might have preferred the elimination of even this road, this position was not realistically advanced at the negotiations. The largest highway proposed was a 10-lane design—8 lanes for cars and trucks, and 2 for buses. This proposal provided an upper bound on the range of possible settlements. All of the disputing politicians were aware that their failure to agree on an acceptable design could result in the loss of federal highway funds, which would provide many jobs and a boost for the local economy. An agreement among the interested and affected town and government agencies would offer great chances for gains over the status quo (Cormick and Patton, 1977).

The existing five-lane road provided three lanes in the peak commuting direction. Negotiations proceeded from this logical starting point to a highway design which offered three general use lanes in each direction and two transit lanes. The negotiators then attempted to incorporate design modifications to make the highway acceptable to the affected towns. Improvements such as depressing and covering the highway through urban areas allowed many groups to realize gains over the status quo.

Despite the fortunate outcome in this situation, no single technique exists for generating a text which individuals will accept as a fair starting point. Generating a text can be especially difficult when a single issue dominates the concerns of both individuals. It is unlikely that either side will even permit a hypothetical modification of their bargaining position. In my view, it is unclear how applicable single-negotiating-text bargaining will be to situations without a natural starting point. On the other hand, a skilled mediator may confer a naturalness on a starting point which does not at first seem natural to the bargainers. In collective bargaining, for example, a mediator may report information concerning settlements in related industries to the negotiators. This information may provide a seemingly natural starting point which could be modified to reflect conditions in this new bargaining context or in the particular industrial setting.

Managing Offer-Counteroffer Bargaining

In offer–counteroffer bargaining, negotiators make reciprocal concessions until they either agree on a settlement point or abandon bargaining. This type of bargaining is probably the one most commonly used (Simkin, 1971). In this form of bargaining, the level of the opening demand and the pattern of concessions probably reveal much information concerning a negotiator's reservation levels and preferences (Pruitt, 1981; Raiffa, 1982).

In offer–counteroffer bargaining, each negotiator will want to signal to the other where he or she hopes to make concessions. Simkin (1971) reports that when labor and management negotiators have a lot of experience, they often engage in confidential discussions prior to the start of formal negotiations. Douglas (1962) reports that informal bargaining sessions can sometimes help to arrange reciprocal concessions. Pruitt (1971) states this is a major function of informal communications. This allows each side to indicate to the other in a confidential way their bargaining expectations, and also where they want to make bargaining gains. Additionally, the confidential prenegotiation communications may help each side to avoid making commitments which destroy the possibility of agreement.

When, however, the negotiators lack much bargaining experience, are meeting for the first time, or are especially hostile to each other, they may fail to undertake these confidential communications. In such situations, it may be difficult for the negotiators to discover which concessions each side would most appreciate. The assistance of a mediator may prove particularly important for establishing a confidential communications channel where none exists. Simkin (1971) states that one of the mediator's tasks may be "to smoke out the priorities" of each negotiator and try to signal these to the other side (p. 99). This information may allow the negotiators to make the concessions which provide the most benefit to the other side. Jackson (1952) states that one function of a mediator is to delineate the intentions of the bargaining opponent. Hearing this information from a mediator can give it added credibility. Kissinger (1982) describes in detail how Anwar Sadat and Golda Meir used his mediation efforts to explicitly describe potential concessions which they would not discuss with each other. In particular, his discussion of shuttle diplomacy (pp. 799–853) describes the specific confidential offers that each side made to him or her for use in the search for settlement.

As in single-negotiating-text bargaining, the intervention of a mediator in give-and-take bargaining may allow the negotiators to make simultaneous bargaining moves. A mediator, in confidence, may explore a contingent concession, either on his or her own initiative or at the suggestion of a negotiator, and announce publicly only when a bargain is made. This screening can limit the possible devaluation of a contingent concession which a formal announcement might produce.

In offer–counteroffer bargaining, a negotiator faces incentives to seek bargaining gains by misrepresenting his or her preferences to either the mediator or the opposing negotiator. The repetition of bargaining, however, provides an incentive to prevent exploitive uses of the mediator. The manipulation of a mediator may cause the victimized bargaining party to refuse mediation in the future and to take other steps to get even. With the check of repeated negotiations, mediators must guard against manipulation that will erode the acceptance of third-party assistance in future disputes.

Conclusion

Real bargaining takes place in a realm of partial information. The scarcity of information concerning the preferences, reservation levels, and even actions of the bargainers creates incentives both for the cooperative exchange of information and for the competitive deception of bargaining opponents. These incentives complicate the bargaining task. In development conflicts, great asymmetries of information are especially likely to exist. Developers will have detailed knowledge about their proposed projects, while a community group may lack both information and the specialized expertise needed to assess it. On the other hand, local groups may possess a concrete understanding of the impacts of a new project on their community, while developers have only abstract conceptions. To promote nonlitigation alternatives for resolving disputes, it will likely prove necessary to enable community groups to develop their own sources of information and data. Although those with a strong adversarial bent may simply use resources to build a case for litigation, the lack of an independent ability to check the veracity of a developer's assertions will likely erode negotiation efforts.

When groups have developed a mature and continuing bargaining relationship, they can establish communications channels that convey important bargaining information while reducing the risk of strategic exploration. Two common information channels include:

- *Tacit communications.* Over time, bargainers develop mechanisms that implicitly signal to each other the firmness of demands and areas for potential concessions.
- *Informal conferences.* At times, bargainers elect to speak off the record to each other and directly discuss the shape of potential agreements. In off-the-record conferences, bargainers may back away from concessions offered but not accepted.

These channels are dominated by norms of truth telling, which are enforced through the concern of bargainers for their reputation. These communications channels require trust between the competing bargainers and are generally available only to those who have established an ongoing bargaining relationship. Once again, developers and community groups will generally lack the continuing relationship needed to initiate productive conferences.

The importance of information revealed through bargaining depends on the information the bargainers already possess. In first-time bargaining, the information revealed by the communications between negotiators will likely have important strategic value, and a mediator's intervention may reduce the chance of its exploitation. When negotiators know each other well, the bargaining probably reveals little new information concerning negotiators' preferences or reservation levels.

Working in a world in which only some things are known, mediators can play important roles in facilitating negotiated settlements through their control of the communications between the disputing parties.

- When bargainers with different negotiating skills or temperaments meet, mediators can faciliate a search for a settlement by physically separating the bargainers and controlling the tone of bargaining exchanges.

- Mediators can facilitate the start of negotiations between bargainers of different skills by screening the weaker negotiator from face-to-face confrontations.
- Even between experienced bargainers, mediators can facilitate the settlement search by screening the source of concession offers. This screening reduces the risks to reputation or bargaining position of making the first move.
- Mediators can overcome distrust by offering bargainers the possibility of making simultaneous bargaining moves. Thus, neither side need wait for a reciprocal concession or need make a concession offer that is either rejected or devalued.
- Mediators can manage the bargaining process. Mediators can prove especially helpful in managing the communications necessary for single-negotiating-text bargaining.

The mediator's actions in controlling bargaining communications may enable the negotiators to emphasize the cooperative elements of the negotiations. When negotiators bargain over many issues, the agreement will have the character of joint problem-solving. Through the exchange of information, negotiators may discover joint concessions which leave both sides better off then the status quo or a noncooperative resolution. This process differs markedly from single-issue bargaining, where one side gains what the other loses.

This examination of information and communications in bargaining suggests that development conflicts face many of the communications problems that arise in international relations, where trust, continuing relations, and a common language are rare. As our analysis has explored methods for overcoming these obstacles, the role of third-party mediators in negotiations has grown. When confronting asymmetries of information and communications obstacles, mediators have managed communications between groups and structured bargaining interactions to promote settlements. Chapter 7 analyzes the even greater role that they can play in using resources to support weak negotiation processes.

CHAPTER 7

Distrust, Limited Resources, and Uncertainty

Introduction

In searching for a constructive resolution to conflict, it is natural to focus on the chief actors in a dispute, the issues that divide them, and the settings in which they interact. Generally, disputes cannot end until some form of resolution is reached between the principal disputing groups on the main issues. Discussing underlying interests and discovering alternatives that offer gains to all depend on having a knowledge of interests and facts that usually only the groups and actors in a dispute possess.

Despite this primary focus on negotiators, mediators have made a forceful appearance in this discussion. In conflicts lacking formal structure, mediators can help to moderate interaction between disputants. Mediators can handle a variety of procedural and logistic tasks that negotiators routinely encounter. Mediators can facilitate communication between disputing parties. In particular, they can enable negotiators to make simultaneous concessions and can permit bargainers to explore bargaining issues without jeopardizing bargaining positions.

The importance of a mediator's work varies with the relationship that exists between the parties. In development disputes, where the participants seldom know each other well, mediators have proved crucial to the achievement of a negotiated settlement, while in labor–management negotiations, mediators appear on the rare occasions when bargainers cannot reconcile their differences. Thus the availability of mediation services can assist all forms of bargaining but helps most when the bargainers have limited experience with each other.

Mediators who possess not only skills but power and resources that they are willing to contribute to a particular negotiation can especially help bargainers to overcome the problems of distrust, limited resources, and uncertainty that plague bargaining. The role of a resourceful mediator is most dramatically seen in international relations and development conflicts, where disputants often distrust each other, have difficulty making credible promises, and face an uncertain future.

President Carter's effort to mediate between Sadat and Begin provides a dramatic example of how a mediator with power and resources can aid disputants to end years of bitter conflicts (Carter, 1982). From September 6 to September 17, 1978, Carter, Begin, and Sadat met at Camp David, Maryland, to negotiate an agreement that would advance the cause of peace in the Middle East. The participants emerged on September 18 to announce a framework for peace (Sobel, 1980). In the joy that followed, Sadat and Begin won the Nobel Peace Prize, but by the time of the award ceremony in December 1978, talks for a final, detailed peace treaty had hopelessly stalled.

Early in the new year, Carter attempted to revive the stalled peace talks, but in February of 1979, both Begin and Sadat refused to accept Carter's offer of a second meeting at Camp David to negotiate a final peace treaty. The expectation of failure and perhaps another war loomed throughout the three countries. In a sharp departure from cautious diplomatic practice, President Carter flew to Jerusalem and Cairo in an attempt to break the deadlock. Carter shuttled between the two cities from March 8 to 13, 1979, in a series of meetings with Sadat and Begin. These meetings led to a peace treaty ending a 31-year state of war between Egypt and Israel (Carter, 1982).

Although these events appear to have sprung from the magic of mediation, the Egypt–Israel negotiations illustrate how skilled bargainers can use the offices of a mediator with power and resources to overcome the distrust, uncertainties, and limited resources that often block the success of bilateral negotiations. Although each negotiation is unique, many disputes share the same bargaining obstacles of distrust, uncertainty, and limited resources, which can deter a search for a negotiated settlement.

This chapter examines how distrust, limited resources, and uncertainty affect bargaining. Surmounting these obstacles is often beyond the power and skills of the bargainers. As the analysis turns to these issues, the focus of this chapter widens to examine the unique contributions that mediators with power and resources can make. This analysis proves particularly relevant both to international relations and to efforts to resolve development disputes through negotiation.

Enforcing Contracts and Supporting Promises

In a world of partial information, uncertainty, and distrust, negotiators may be unable to effectively make promises to each other. When there is no formal legal mechanism to force a bargainer to honor commitments to future actions—as often occurs in international negotiations and in some environmental-development disputes—then the inability to make an effective promise can prevent negotiators from reaching a settlement.

Schelling (1960) discusses in great detail the strategic value of a promise. He describes a promise as a commitment to future action. A promise is necessary "whenever an agreement leaves any incentive to cheat" (p. 43). Promises enable negotiators to search not only for agreements conditioned on simultaneous exchange, but for agreements that require future responses for present actions. With promises, bargaining can search for settlements not only in the present but also in the future.

Promises are important not only in resolving formal disputes, but in many daily activities. Modern commerce relies on the ability to make legal contracts. Without the ability to commit themselves to future actions, such as the delivery of goods and services, or the payment of a debt, businesspeople would either rely entirely on cash-and-carry transactions or else organize along ethnic, religious, or family lines, where social rather than legal sanctions would guarantee contracts. When bargainers cannot make promises, the range of possible settlements becomes limited in time, and bargaining resolutions are reduced to those exchanges that can take place simultaneously, as in a simple bazaar or barter economy.

Despite the weaknesses of a promise when society's structures fail to guarantee them, a promise, even when backed only by social norms, may make an offer acceptable to an opposing bargainer. Concerns for reputation, whether for business, diplomatic, or personal reasons, can prove an effective means of giving credibility to promises. Bookies pay off winning gamblers despite the absence of legal sanctions. Failing to do so would threaten their generally lucrative business. Within a family, promises are often backed by moral norms and the threat of ostracism, which can carry even more force than the sanctions of the state.

No promise, however, is a certain guarantee of future action. Any experienced negotiator will make an assessment of the likelihood that a person or nation will honor its promises. A negotiator's acceptance of an agreement that relies on promises will depend on a balancing of both the risks of falling victim to a broken promise and the gains offered by an honored agreement against the consequences of nonagreement.

When institutions do not exist that can enforce promises, some arrangements can arise to solve these problems. Although the illegality of betting prevents the use of legal contracts to guarantee debt collection, people betting can use a third party whom both trust to hold their stakes (Schelling, 1960). This arrangement can greatly reduce the problems of collecting a gambling debt. In other illegal activities such as drug trafficking or loan sharking, organized crime can act as a shadow government, guaranteeing transactions and debt collection among its members.

Labor-management contracts are legally binding. International relations, however, lack both the institutional mechanisms for enforcing contracts and the social or religious norms that help guarantee promises. In relations between nations, there does not exist a higher authority which can support the promises and the treaties of a country. Reputation alone serves as only a minor deterrent to treaty violations. Iklé (1964) comments that dictators have often violated even the basic diplomatic norm of not harming a negotiator, but he remarks ironically that people and states have short memories in these affairs. He recounts that Stalin imprisoned Polish leaders whom he invited to Moscow in 1945 to form a new Polish government. They were tried in Russia and sentenced to long prison terms. One would think that this trick could work only once. However, following the Hungarian uprising in 1956, Stalin invited Hungarian leaders to Moscow to complete agreements for Soviet troop withdrawals. He imprisoned and executed them. More recently, the seizure of American diplomats in Iran in 1979 showed the inability of the international community to make a dictator conform to even this basic diplomatic norm.

Negotiators may use a mediator with power to enhance their ability to make promises. In particular, a powerful third party may offer to support the terms of a negotiated treaty by pledging either to take actions against the party which violates the agreement, or to compensate the wronged party. The pledges of a powerful mediator may both change a negotiator's assessment of the likelihood that a bargaining group will keep its word and decrease the size of the potential loss that a negotiator may sustain from a broken promise. Both of these products of a mediator's intervention and pledged actions may alter a negotiator's risk-benefit calculus to enable him or her to accept agreements with promises from a bargaining opponent whom he or she could not otherwise prudently trust.

In the search for a Middle East peace, the United States has often acted as a powerful mediator, helping to guarantee treaties negotiated between Egypt and Israel following the October 1973 war. The United States' role as a mediator began after Egyptian President Sadat expressed a willingness to negotiate a peace under U.S. auspices. Initial U.S. actions to act as a fair and trustworthy mediator helped to insure Egypt's acceptance of the United States. Kissinger (1982) relates that after a ceasefire had been arranged between Egypt and Israel, Israel violated this ceasefire to entrap Egypt's Third Army in the Sinai Desert. Kissinger then sent a United States message to Israel stating that the United States would not accept "the destruction of the Egyptian army under conditions achieved after a ceasefire was reached in part by negotiations in which we participated" (p. 608) and threatened to support United Nations action to enforce the ceasefire resolution. In response, Israel agreed to permit the resupplying of this army with food and water. In my view, this incident helped build Sadat's faith that the United States could act as a fair mediator and enforce future negotiated promises.

Subsequent shuttle diplomacy helped produce a disengagement of Egyptian and Israeli forces in 1974. Under the terms of this agreement, United Nations peacekeeping forces took up positions between the opposing armies to inspect the ceasefire and limitations of forces in the Sinai (Kissinger, 1982), thereby helping to guarantee and enforce the promises of both sides to halt military actions.

The negotiations leading to the Sinai Accord of 1975 offer another particularly clear example of how a powerful mediator can enforce promises when no institutional enforcement mechanism exists. Following the ceasefire, negotiations faced a major bargaining obstacle. To Israel, the Sinai Desert represented a buffer between it and a potentially hostile and powerful Arab state. Withdrawing beyond the strategic Gidi and Mitla passes increased Israel's vulnerability to sneak attack. No promise by Egypt could offer a security that equaled the military possession of these passes.

The mediation efforts of the United States in this second round of negotiations enabled both sides to accept the other's promises of nonbelligerence. In this second Sinai agreement, United States civilians, stationed in the Gidi and Mitla passes between the opposing troops, monitored the demilitarized zone with electronic gear. Violations of a ceasefire—and, in particular, a frontal assault—would virtually require that the aggressor nation overrun the American camps. In effect, U.S. civilians served as hostages to the agreement and were probably a more powerful deterrent than the United Nations troops. Any harm to those civilians carried an implicit threat of U.S. military action against the aggressor nation. Their presence insured both that the United States would be immediately involved in any new hostilities and also that domestic sentiment in the United States would mount against the aggressor. This placement of U.S. civilians pledged the United States to enforce the separation-of-forces agreement perhaps even more effectively than a diplomatic pledge.

Throughout the entire Egyptian-Israeli search for peace, the United States continued to play the role of a powerful mediator guaranteeing promises. Even at the signing of a peace agreement in March 1979, the U.S. government signed a memorandum of agreement with the State of Israel "to take appropriate measures to promote full observance of the Treaty of Peace." Only the involvement of the United States permitted both sides to overcome the bitterness and distrust bred by war that would have made promises of peace unbelievable.

Subsidizing Settlements

At times, negotiators may bargain earnestly only to discover that there are no possible settlements which they can both accept. Even the intervention of a mediator cannot always help. A mediator may find that he or she cannot persuade the negotiators to either change their preferences or back away from commitments that preclude settlement, or that his or her guarantees of bargaining promises prove inadequate to permit acceptance by the negotiators.

In cases where the bargaining agenda or interests are too limited to support a negotiated settlement, a mediator may elect to subsidize a settlement or make available to the negotiators potential settlements that they could not reach alone. This action is particularly appropriate when the mediator has a strong interest in seeing the disputants reach a negotiated settlement on their own. Even if not a direct participant in a dispute, a mediator may have a strong interest in seeing that the negotiators reach a peaceful and negotiated settlement. With the approval of the bargainers, mediation may offer the third party a mechanism for protecting its own interests while it serves the interests of the disputing parties.

The United States played such a third-party role in the Camp David peace process. To support the final settlement, the United States not only offered to guarantee treaty terms but gave aid to Egypt to help it to develop its economy and to Israel to help it build air bases to replace those left in the desert. This aid helped Egypt accept the disapproval of other Arab states without jeopardizing its own economy and permitted Israel to return occupied land to Egypt without seriously jeopardizing its security. These subsidies may not only have helped make an agreement possible but may also have constituted a wise investment by the United States. United States interests, as well as those of Egypt and Israel, are well served by the avoidance of war between nations in the Middle East.

In disputes over the siting and construction of development projects, rarely do we see a powerful mediator intervene to open up new potential settlements. The resolution of a dispute over the use of federal land in the Gospel-Hump region of Idaho, however, offers a rare example of a powerful mediator offering disputants in a development conflict access to settlements which they could not generate themselves (*Congressional Record*, 1977). In this dispute, timber interests and local environmentalists fought over the use of national forest land in a remote region of the state. The dispute between these two groups had delayed any decision and clouded the final disposition with uncertainty. Senator Frank Church's position on the key Senate Committee on Energy and Natural Resources, however, allowed him to offer to pass legislation to support any plan that carried the endorsement of the representatives of the timber industry and local environmentalists. His offer gave these groups the chance to resolve their differences through negotiations and access to settlements not otherwise available to them. In particular, his offer of legislation allowed the negotiators to avoid the protracted administrative hearings and reviews that the Forest Service must perform to ratify any land-use plan. This review process would have required legal expenditures by all participants. Further, a required moratorium on timber harvesting during this lengthy review would have produced a shortage of timber and created severe hardships for the local economy. Senator Church's endorsement of a negotiation process and his offer to pass a law implementing the terms of agreement made a settlement possible on terms unavailable without his aid.

Designing Agreements to Generate Trust and Overcome Uncertainties

Uncertainty over the future actions of bargainers or over the outcome of events beyond the control of negotiators can impede the negotiated resolution of a dispute. When a common perception of future uncertainties limits negotiations, skilled negotiators and mediators can design agreements that develop trust sufficient to permit them to accept each other's pledges. Similarly, when negotiators hold different perceptions of the likelihood of future events, skilled negotiators and mediators may design agreements that exploit these different perceptions. A common technique used to resolve lack of trust between bargaining opponents is to divide the problem of resolving a dispute into a series of smaller tasks. The sequential accomplishment of these tasks can help the negotiators to generate the trust necessary for the acceptance of a final agreement, while reducing the risk that either side takes at any given step in the negotiation process.

As described earlier, in international negotiations, diplomats often negotiate a series of treaties dealing with the same issue, such as the treaties between the United States and the Soviet Union to limit the development and use of nuclear weapons. After the terms of one treaty are fulfilled, representatives of the countries can meet to negotiate a new treaty that can address additional issues or respond to emerging conditions. Such a process can help the countries to make progress toward a long-range goal.

Implementing a treaty in stages offers another way of generating trust. Over the course of the treaty, each step remains contingent on the fulfillment of the conditions of the earlier stages. Troop withdrawals and military disengagements often require the establishment of such a series of staged reciprocal actions. Although these forms of settlement can help nations overcome doubts over whether groups will honor a treaty, this process also generates trust that spills over into ongoing relations.

The Egyptian-Israeli negotiations from 1973 to 1979 used both sequential treaties and the sequential implementation of treaty provisions to generate confidence in the search for peace. Following a negotiated ceasefire concluding the 1973 war, a series of treaties negotiated over the next six years led to the signing of a peace treaty in 1979. These negotiations included:

- The Sinai I Accord of June 1974, which arranged for a simple separation of forces;
- The Sinai II Accord of September 1975, which arranged for a withdrawal of Israeli forces and a U.S. civilian presence to monitor treaty violations and troop movements;
- The Camp David Accord of September 1978, which developed a broad framework for peace; and finally,
- The Egyptian-Israeli Peace Treaty of March 1979, which required a phased withdrawal of Israeli troops and an exchange of ambassadors between Egypt and Israel.

In retrospect, we can easily see how the representatives of Egypt, Israel, and the United States used a series of treaties implemented incrementally to establish a process that helped to generate the trust between the two countries necessary for a lasting peace. One treaty deferred unresolved issues to the next phase of negotiations, and the parties implemented treaties as a series of reciprocal steps that helped generate the trust needed to take still greater risks for peace. Even the final peace treaty left issues unresolved. The difficult issue of determining a mechanism for protecting the rights of the Palestinians was deferred to some future round of negotiations, and this issue still remains unsettled.

Such a sequence of treaties and actions may also help opposing countries overcome their concern that a country will fail to comply with the terms of a treaty. At each step, noncompliance may cause only an inconvenience to the wronged nation. Thus, small risks are taken before large ones, and good relations reduce the perceptions of the absolute level of risks. If all terms of an agreement are implemented at once, risks may seem large and a treaty violation may compromise the security of the wronged nation.

All treaties serve a domestic function (Iklé, 1964). Dividing the negotiations and implementation of a treaty into a series of small steps may enable the countries to overcome constraints posed by domestic politics. If the people of two nations hate each other, then only time and compliance with early treaties may permit the people to alter their opinions and allow the nations to eventually accept a total resolution of the conflict. Kissinger (1982) suggests that just such a process of readjustment of perceptions concerning the opposing nation was necessary to permit an Egyptian-Israeli accord. He advised Sadat to approach negotiations with Israel as a psychological problem, rather than as a diplomatic one. Similarly, Israel had to change its perceptions of Egypt's military capabilities, and its will and desire for lasting peace.

Only new information and the passage of time can resolve the uncertainties over future events. When bargainers anticipate different futures and cannot reach an agreement, then new information can sometimes reduce uncertainties. Over time, events may create new facts. By spreading negotiations over time, bargainers both establish a bargaining relation and allow the unfolding of events to create possibilities for new settlements. The existing apparatus for negotiation can then enable bargaining nations to capitalize on new developments.

Even when negotiators cannot wait for or predict the future, experienced bargainers can identify the important issues in any negotiating relationship. When the consequences of present action are not irreversible, bargainers can plan for future events either by agreeing to reopen negotiations, or by negotiating settlements which include clauses that call for actions contingent on the outcome of future events. In the collective-bargaining relationship, both forms of settlements are readily observed. Many contracts call for periodic reopening to discuss developments that alter labor–management relations. Other contracts call for cost-of-living adjustments automatically tied to the consumer price index.

In development conflicts, only rarely do the opposing groups manage to develop an ongoing relationship that they can maintain through an uncertain future. Thus, problems associated with the enforcement of an agreement can loom particularly large. Not only is it difficult to develop legally binding agreements, but the lack of a continuing relationship linking the conflicting parties limits the possibility of reaching an agreement that can respond to future developments.

Designing Alternative Forms of Risk Bearing

Uncertain future events impose an element of risk on negotiations. Different sides in a dispute, or even individuals on the same side, may have very different attitudes toward these risks. It is possible for two persons to agree that a course of action offers them good prospects, yet one may fear that he or she cannot accept the chance that a future event may turn against his or her interests. This fear may cause the negotiator to prefer the certain prospects offered by no action to the possibly beneficial prospects offered by a risky course of action. On the other hand, not making a decision may also pose risks. In a development conflict, failure to negotiate invariably leads to litigation. The outcome of a court case may be highly uncertain, and both sides may prefer to negotiate a resolution rather then risk the possibility of a court decision that will go against their interests. Thus, both sides may desire to negotiate in order to control risks.

When both negotiating parties fear the risks that the future holds, they may find that they cannot develop any agreement that overcomes their aversion to the risks in a new agreement. In such a situation, without the intervention of some outside agent or change in perceptions, the negotiators will find that they cannot negotiate a settlement. A mediator that has sufficient resources and a strong interest in a dispute can, however, offer to insure an agreement against uncertain future events. Just as enforcing promises can help produce agreements, insuring against risks can enable bargainers to accept agreements that they otherwise could not.

Once again the peace treaty between Egypt and Israel offers a clear example of how an interested mediator can help parties reach agreement by offering to insure against uncertain future events. As part of this treaty, Israel returned to Egypt oil fields captured in the 1967 Mideast war. Israel feared that the loss of these oil fields would leave it dependent on uncertain supplies of oil imports. In a memorandum of understanding, the United States offered to guarantee Israel a supply of oil at world market rates for 15 years if Israel faced a disruption of oil supplies. Once again, the combination of United States resources, domestic oil supplies, and a commitment to this peace treaty made it possible for the United States to insure Israel against this future uncertainty. To date, the United States has not needed to supply Israel with oil. Thus, the insurance has cost the United States nothing.

Negotiating parties may have quite different attitudes toward risks and different perceptions concerning future events. Just as asymmetries of preferences can permit bargainers to realize joint gains, asymmetries in attitudes toward risks or varied perceptions of the future can permit bargainers to design agreements that incorporate contingent clauses that permit joint gains.

A mediator skilled in negotiations can assist inexperienced bargainers in designing agreements that effectively exploit these differences in perceptions. This technique can prove particularly important in negotiations between sponsors of development projects and those that oppose them. Developers often see risks as opportunities and can have a difficult time understanding the fears of project opponents who dread risks and see the future as harboring threatening events (Susskind et al., 1978).

In a dispute over the construction of a shopping mall in White Flint, Maryland, a suburb of Washington, D.C., local groups opposed the shopping center for a variety of reasons, including their fears that the mall would have an adverse impact on the property values of homes abutting the shopping center (Rivkin, 1977). Although the shopping center planners incorporated many design features such as the construction of a berm separating the center from the neighborhood and agreed to prohibit the use of the mall's parking lots for fund-raising events such as fairs, there still remained a risk that normal commercial activity would lower property values. Such a fear had recently produced active citizen opposition and rejection of an alternative site by the county planning commission. In the final negotiated agreement, the shopping center developers agreed to compensate the owners of property adjacent to the shopping center for any decrease in property values which they suffered during the first five years of mall operation. Through this agreement, the shopping center developers agreed to bear risks that the property owners did not wish to face and which the developers thought were nonexistent. This transfer of risks, at no present or expected cost to the developers, from the homeowners to the developers enabled them to win the assent of the local residents and avoided a prolonged and potentially endless debate over whether a reduction in property values would occur.

Conclusion

In disputes over development projects, negotiators often find that they are unable to develop a mechanism for making enforceable promises to each other. Unless the individual groups have the power to make legally enforceable agreements, they may find that they must bargain without the ability to make promises. This restriction seriously limits the ability of any bargainers to reach a negotiated settlement.

Distrust, limited resources, and uncertainty can pose obstacles that tax bargaining skills and often require the assistance of third parties. Acting alone, skilled negotiators can:

- Develop a negotiation process that uses incremental steps to develop trust and bargaining momentum;
- Design agreements that address anticipated uncertainties or future developments;
- Develop risk-sharing agreements that use diverging perceptions and attitudes to develop settlements that generate joint gains.

Although these techniques offer a poor substitute for the institutional endorsement of law, such techniques can at times prove useful.

A mediator who brings substantial resources or authority to a conflict may act to change both the nature of the bargaining situation and the range of possible outcomes in a variety of ways that enhance the possibility of a settlement. The mediator can:

- *Guarantee promises.* A powerful mediator may offer to guarantee the promises of the bargainers. When a mediator with resources agrees to enforce the settlement terms or to compensate those wronged by a violation of an agreement, then settlements that rely on promises can prove attractive to disputants.
- *Subsidize settlements.* A powerful mediator can use resources to expand the range of possible outcomes by directly subsidizing settlements.
- *Insure against adverse outcomes.* A powerful mediator can assist bargainers by offering to insure agreements against unforeseen or particularly adverse future events beyond the control of the negotiators.
- *Design contingent agreements.* Armed only with skill, a mediator may prove helpful to bargainers who cannot surmount the anxiety sur-

rounding uncertain future events. A mediator may help the disputants design agreements that require action in response to future events.

- *Explicitly examine uncertainty.* A mediator may be able to recast the bargaining situation into a framework that allows the negotiators to incorporate uncertainties explicitly into the terms of their settlement.
- *Exploit discrepancies in risk preference.* A meditator may be able to suggest risk-bearing arrangements that permit the negotiators to use their different attitudes toward future risks to aid their search for a settlement.

These actions of a mediator can prove most helpful when institutional settings fail to support negotiated settlements. In both international relations and development conflicts, interventions by resourceful third parties can particularly enhance the ability of bargainers to find acceptable settlements.

CHAPTER 8

The Exercise of Negotiator
and Mediator Discretion

Introduction

This book began with a description of the problems generated by
the bureaucratic and adversarial processes now used to review proposed
development projects. Even when a project produces net benefits, a local
community will often bear costs that outweigh its share of total benefits
(O'Hare, 1977). When communities and individuals express opposition
through administrative and judicial challenges, their actions can block
needed and beneficial projects. Often opponents of a project challenge
the adequacy of an environmental impact statement (Bardach and Pugliaresi, 1977; Frieden, 1979) and demand a fuller consideration of alternative sites. Developers make decisions early, announce them to communities and regulatory bodies, and then struggle with regulatory and
judicial reviews (Ducsik, 1978).

Few would doubt that development and administrative decisions
distribute benefits and costs to communities, individuals, firms, and people at large. Thus, siting decisions are not decisions that affect only the
interests of those making them, but decisions that shape the character of
a community and its environment and affect many individuals not
directly participating in the decisions. The evolution of administrative
law and regulation has created procedures that identify and highlight
the choices and value judgments that underlie these decisions. Environmental laws that require the identification and assessment of the effects
of projects on a locality or region help generate information concerning
the likely effects of a project. These laws probably discourage projects

that clearly produce great social harms. Many projects, however, fall into a murky middle ground in which the decision to proceed depends greatly on the weights placed on key impacts. Unfortunately, laws and court interpretations do not provide much guidance to the decision maker who must weigh this information.

The current complex regulatory and administrative proceedings did not grow out of a collective irrational impulse to halt beneficial projects. Rather, these proceedings developed both to protect communities against the effects of inappropriate facilities (Bardach and Kagan, 1982) and to check the exercise of discretion by bureaucratic agencies empowered to safeguard the public interest (Friedman, 1979; Stewart, 1975). Laws, institutions, and procedures evolved to protect individuals and communities against the adverse effects of decisions, whether the construction decisions of developers, or the permit decisions of government administrators.

This book argues that a wider use of negotiation and mediation could help transform this decision-making process from a zero-sum contest into a collective search to determine whether some development package could prove more acceptable to all. The negotiation process could induce developers to incorporate design changes to mitigate adverse impacts and to seek ways to compensate those who bear costs, while offering developers more certain and speedy decisions. Negotiation and mediation procedures explicitly confront the public choice dimension of development decisions. Negotiation processes recognize that direct public participation in siting decisions may produce quicker and better decisions that are viewed as fair and legitimate by all.[1]

Negotiation and mediation processes work by supplementing the traditional discretion of the bureaucratic or administrative decision maker with the discretion of the bargaining participants. It is the discretion that the negotiators and mediators exercise in negotiations that can enable them to search for agreements that leave all better off. Furthermore, the negotiation process permits individuals to use discretion and common sense to resolve the diverse issues resulting from local conditions and project details. Negotiated agreements can offer both a process that is more popular and an outcome that produces fewer social costs than the decisions produced by adversarial processes.

---

[1]In a major study of the Army Corps of Engineers, Mazmanian and Nienaber (1979) found that when the Army Corps developed alternative project designs and consulted with representatives of local communities, the respondents gave high marks to the planning process but responded less favorably to the final proposals.

Although one can paint an optimistic picture of the outcome of negotiations, it is equally easy to imagine a process in which individuals abuse discretion and the public trust. If corrupt negotiators replace honest bureaucrats, then the community may discover that honest developers find their projects delayed while less scrupulous builders bribe negotiators. Graft may enable unsafe or harmful projects to gain approval. Even when corruption does not arise, conscientious citizens will wish to insure that negotiation processes will not permit local communities to subvert larger regional or national goals, such as those for a cleaner environment, fair employment practices, or secure energy sources. Others will correctly worry that negotiations may fail to represent the interests and concerns of some minority group or may sacrifice its concerns in return for gains realized by a majority coalition. Negotiation, instead of making all or most better off, may make some better off while making others much worse off.

The exercise of discretion by negotiators and mediators in disputes over development projects is special in several ways. Unlike in industrial relations, law generally fails to endorse the negotiation effort or to limit the discussions to issues and solutions that society finds acceptable. Similarly, law seldom holds bargainers accountable to the groups they represent, and no government agency polices the actions of mediators. The consequences of a settlement may fall not only on the negotiation participants, but also on many unrepresented in the bargaining. Although this can also be true in industrial relations, market competition and foreign imports limit the ability of labor and management to pass the costs of a contract onto consumers, and Taft-Hartley injunctions can enable government to limit the consequences of strikes. Unlike either industrial or international relations, bargainers in development disputes may lack the skills and powers that enable them to adequately represent their interests. This lack increases the discretionary power of mediators. The one-time nature of many development disputes further weakens the value of reputation in checking abuses of discretion.

This chapter examines ethical and political questions raised in the use of a negotiation process to resolve development disputes. It explores issues that arise in the selection of an issue agenda and participating groups. Further, it addresses the classic problems raised by representative bargaining—how to prevent corruption of a group's leadership and how to protect the interests of minorities from a tyranny of the majority. Finally, it explores the duties of negotiators and mediators to each other, to the bargaining process, and to society at large. This analysis will suggest ways of bounding the range of discretion in negotiations, of holding bargaining participants accountable for their actions, and of preventing and penalizing blatant abuses by negotiators and mediators.

Issue Agenda and the Participation of Groups

The choice of issues for a bargaining agenda and the selection of groups as bargaining participants greatly affect the final disposition of any dispute. Earlier, we saw that a rich issue agenda can enhance the ability of disputants to reach a settlement by permitting a wider search for agreement and by creating opportunities for reciprocal concessions that produce gains for all. Most practically, however, the selection of an agenda determines what the bargainers can discuss, as well as the character of the final settlement.

Similarly, those who participate in a bargaining session will influence the course of negotiations and the final outcome of bargaining. In particular, the selection of the bargaining participants determines which groups get represented, and which interests are heard at the bargaining table. Thus, the recognition of groups as legitimate participants in bargaining not only serves to help initiate negotiations but also helps determine whether a settlement is possible, and what its final shape will be.

In collective bargaining, laws determine which issues are appropriate for inclusion in bargaining, and which groups are legitimate bargaining participants. Furthermore, the Landrum-Griffin Act outlaws corrupt practices such as the direct compensation of negotiators (bribes) or a payment for labor peace. This law seeks to rule these issues off the bargaining agenda and formally reinforces the widespread opinion that these concerns are not legitimate bargaining issues.

The National Labor Relations Act also sets a procedure that determines who is a legitimate bargaining participant. Such a policy not only prevents the violence that was historically associated with disputes over union recognition but also plays a large role in determining the shape of the final agreement. In particular, the determination of the size and constitution of the appropriate bargaining unit can determine the bargaining demands. Large industrial unions, for example, must balance the demands of skilled workers, who often constitute a small percentage of the workforce, against those of unskilled workers, who often make up the greatest proportion of a workforce. Government action removes the decision over the scope of a bargaining unit from the hands of either industry or labor.

In international relations, representatives of nations spend much time in prenegotiation conferences bargaining over what issues should go into a formal bargaining agenda, and which nations can participate in the bargaining sessions. With no institutional means for resolving these issues, the prenegotiation bargaining, often conducted through intermediaries, can lead to complicated and prolonged discussions in preparation for bargaining.

In environmental-development negotiations, setting the issue agenda and selecting the bargaining participants can lead to controversy. In particular, since the law generally does not provide guidelines concerning what are legitimate bargaining issues, the expansion of the agenda beyond the issue that gave rise to the dispute can lead to more challenges and objections. In addition, since laws often fail to legitimate the discussion of a wide issue agenda, these efforts to discuss other issues can be characterized as an extortion by external forces, or as a sellout by a group's members.[2] The Massachusetts Hazardous Waste Facility Siting Act offers a significant departure in that it explicitly calls for compensation of the local community by a facility developer and lists appropriate forms of compensation, including "direct monetary payments to the town and provisions to insure the health and safety. . . of the host community and its citizens."

Perhaps because most negotiations over development projects have been voluntary and ad hoc, mediators have played an extremely important role, both in initiating negotiations, and in inviting groups or representatives of involved interests to take part in negotiations. To date, mediators and proponents of environmental mediation have argued that any individual or group who desires should participate in negotiations. In part, this is the only practical solution to this issue, since all individuals retain the right to initiate legal challenges to any decision. Unfortunately, this technique can at times lead to negotiations that include large numbers of individuals and groups and can create sessions that more closely resemble public hearings than bargaining sessions. Mediators, as third parties to the dispute, face a difficult if not impossible task if they attempt to limit participation in negotiations. Lacking criteria to guide decisions to exclude groups or individuals, the safest course is to include all.

---

[2]See Susskind (1981) for a discussion of how the judge who reviewed the negotiated settlement of a dispute over the Foothills Water Treatment Project in Colorado both refused to sign the consent agreement and questioned the payment of legal fees to the lawyers representing the opponents of the treatment plant.

Negotiator Discretion

In bargaining, not only do negotiators search for new agreements that leave the group better off, they must often judge whether or not this actually happens. How can one decide whether an agreement actually improves the position of the group either over the status quo or over what they might gain through an adversarial process? In a development dispute, this will rarely prove easy to do. This determination is fundamentally a judgment call and requires the exercise of discretion by the bargaining representative. When the search for agreement requires compromise, then the bargaining representative may find that he or she must balance the competing concerns of subgroups or factions present within the larger group. The negotiator must also weigh the certainty of the negotiated settlement against the uncertain future outcome of litigation. What legitimates these decisions? What checks operate on the discretionary decisions of the negotiators to guarantee that they act in ways that legitimately protect and represent the interests of their group's members?

This question has no simple answer. Fundamentally, a collective decision raises the possibility that it will limit the freedom of some individuals within a group or community (Olson, 1973). Decisions regarding the location of roads, hospitals, and so on, inevitably raise questions concerning what is an appropriate decision or action: When can a group's leadership legitimately compromise the interests and concerns of some individuals for the good of the group? When can leaders legitimately negotiate settlements that require some to suffer inconvenience for the benefit of others?

These issues concerning the legitimate use of authority, the limits placed on individual action, and the losses to some that generate benefits for others constitute central questions of political philosophy and social theory. The debate over what constitutes the just exercise of authority goes back in Western society at least to Plato and continues today in the works of such disparate writers as Rawls (1971) and Nozick (1974). In the United States' representative democracy, legitimacy flows from the people, who are sovereign, to the laws enacted either by their representatives or by direct citizen initiative (Hamilton et al., 1961). The exercise of discretion by government is itself limited by the Constitution. Through elected officials, laws, and judicial decisions, our society strikes balances between the competing values, interests, and desires of disparate groups of citizens. The potential abuses of discretion are limited by formal procedures, by the review of the courts, and by citizen action at the ballot box.

In labor–management collective bargaining, the representatives of a union bargain on behalf of all workers covered by their bargaining unit. These bargainers not only represent the interests of workers but must transform these interests into concrete bargaining demands and must strike compromises between the competing concerns of the members of a union. This need to achieve a balance between the competing concerns of citizens raises smaller but similar issues of legitimacy and discretion. What gives a union bargainer the right to speak on behalf of *all* workers? What forces act to limit his or her exercise of discretion? What insures that the rights of individual workers or job applicants will not be sacrificed to the desires of a powerful coalition?

Although unions emerged with strong ideals of worker solidarity, as the institutional character and the strength of unions has grown over time, the interests of workers belonging to the same union have diverged from the common initial concerns for union power and survival. Younger workers care relatively more about the hourly wage rate, while older workers have concerns for job security and pension benefits; representatives of the national union have concerns for pay scales throughout the industry, while the members of a local may care more about the survival of an individual plant. Although all workers will agree that they desire a larger compensation package, they may have a difficult time reaching a consensus on the division of the compensation between wages, health benefits, and pension investments.

In practice, government regulates the structure of labor unions, forcing them to adopt a system of union democracy that parallels that of a political democracy. The country has solved this question of controlling the discretion of bargaining representatives by using the same methods used in the larger democracy. The Landrum-Griffin Act, in particular, requires that labor unions hold periodic secret-ballot elections that may be supervised by the government. The practices of labor unions are regulated to insure that union leadership will not abuse its power by discriminating against individual workers or by racketeering. Furthermore, not only can workers replace leadership, but they retain the right to decertify a union, thereby insuring that it will remain responsive to the interests of the majority of workers. Once again, in labor–management relations, one can see the intervention of government to both legitimate and regulate union democracy. Union leadership, elected by the secret ballot of members, bargains on behalf of all members subject to the limits of law, and to the political limits of the union ballot box.

In environmental-development disputes, the intervening groups can lack a representative organizational structure. In general, the lack of any procedures to hold leaders accountable for their decisions leaves bargaining actions open to question. Thus, a negotiator's decisions that compromise the interest of subgroups on behalf of the entire group pose problems for the negotiation process. What principles legitimate these actions? What constraints limit the exercise of leadership discretion to acceptable bounds?

Although the post hoc review of negotiated agreements by the courts can prevent abuses of discretion and confer legitimacy on the negotiation process, a negotiation process that requires much judicial supervision offers only modest gains over an adversarial administrative process that relies heavily on judicial review. Such a negotiation process may prove little more than another delaying tactic on a road that leads to litigation.

In my view, a negotiation or mediation review process would benefit greatly from a formal resolution of the legitimate range of discretion by law. Although the endorsement of mediation efforts by key political figures has helped to legitimate specific negotiation attempts (Cormick and McCarthy, 1974; Cormick and Patton, 1977), only formal procedures that clearly address these issues can produce a stable and workable negotiation process. Subsequent judicial review limited to whether the negotiated settlement falls within the limits set by law can provide the necessary protection for individual rights without greatly increasing the uncertainty that bargainers face.

The Importance of Mediator Discretion

Just as important as the exercise of discretion by the representatives of bargaining groups is the exercise of discretion by a mediator. A mediator plays an important role in promoting the success of a negotiation process, and in shaping the final settlement package. A biased mediator could use his or her access to confidential information to promote the interests of a particular negotiating party. A mediator's access to privileged information and his or her ability to control, limit, or modify bargaining communications can have a major influence on the course of negotiations. To function effectively, the mediator requires access to much confidential information. In particular, negotiators will often reveal which deals are acceptable to the negotiating parties, which concessions a bargaining group may either make or desire, and which internal disputes divide a group. Access to this information can enable a biased mediator to direct the course of negotiations against a group so that they receive the minimal reciprocal concession for their bargaining offer. In addition, a biased mediator could press for an agreement that would lie close to a group's reservation level. Such knowledge could permit an unscrupulous mediator to assist a bargaining party to gain the best possible settlement. A biased mediator can therefore pose a serious threat to a negotiation process.

A mediator who takes a particularly active role in the negotiations can find that a bargaining group will try to manipulate his or her actions to advance their interests. In particular, when a mediator actively suggests bargaining compromises or settlements, bargainers may misrepresent their group's real interests or bargaining concerns so that the mediator's proposals will lead to outcomes more favorable to them (Simkin, 1971). If a negotiator feels that a mediator will suggest some compromise that splits the difference between the negotiators on some issue, then the bargainers may seek to overstate the true minimum that they will accept. Communications with the mediator will soon lack either the truthfulness or the openness that is often necessary for success. When negotiators act in this way, the mediator becomes one more element in the strategic bargaining, rather than a true confidante.

To preserve the integrity of mediation, a mediator must take steps to detect such manipulation or to limit its effects on the negotiations. Simkin (1971) advises mediators in collective bargaining to limit their suggestions to rare occasions and to accept a strike rather than rushing in to settle a particular dispute. Decreasing the chance that a mediator will make a proposal can limit the value to the bargainers of misrepresenting their bargaining positions. If a mediator suspects that bargainers are attempting to manipulate his or her actions by misrepresenting their true demands, then the mediator can refrain from acting on this information, either by limiting his or her role to procedural actions or by withdrawing from the negotiations.

As described earlier, mediators make a contribution to a negotiation by screening the sources of suggested negotiation compromises and the confidential communications of bargainers. Such actions can enable negotiators to make concessions without fearing that an offer will signify a bargaining weakness or produce a devaluation of a particular concession (Pruitt, 1971). Similarly, a mediator can use the confidential disclosures of bargainers to direct negotiations to areas where bargaining progress is possible. The ability of a mediator to perform this function depends on his or her ability to protect confidential disclosures. This depends on an ability to prevent bargainers from seeing through him or her to the real source of a bargaining offer or suggestion. A mediator's failure to protect this information, through either duplicity or error, can seriously injure the bargaining position of the groups that confide in him or her. Such failure can undermine both the bargaining effort and the bargaining process.

A mediator can also control the timing of the release of information. To reach an agreement, the bargaining parties must at some point develop an understanding of what terms of agreement will prove acceptable to their bargaining opponents. A mediator can, with the assent of the bargainers, control the release of information to promote progress on specific issues. However, if a mediator reveals information prematurely, then a negotiator may suffer a major erosion in his or her bargaining position.

In the last chapter, I discussed how a third party with power and resources could act to support promises, subsidize settlements, insure against uncertain future events, and suggest arrangements for sharing or bearing risks. When mediators with power and resources intervene in bargaining, the mediator acts almost as another bargainer. In many situations, mediators will use their resources to assist the negotiators because they will also benefit from an agreement. When a mediator has both power and an interest in a settlement, an issue may easily arise over what actions infringe on the autonomy of the negotiators. A third-party participant may use power not only to support a settlement, but to impose a specific resolution on the disputing parties. Once again, it is difficult to see what limits these discretionary actions.

Although discretionary actions by mediators are often necessary for helping the negotiating parties to reach an agreement, they raise several issues:

- Mediator actions, whether through overt bias or indiscretion, can seriously injure the position of groups that confide in the mediator. What mechanisms check the potential abuses of the negotiation process by the mediator?
- Mediators can use power and resources not only to help support agreements, but also to impose them. What principle can guide a mediator in determining which interventions are appropriate and which are not?

Checks on Mediator Discretion

There are several types of checks that act on a mediator to limit inappropriate uses of discretion. The strongest checks exist in industrial relations. These checks include those that arise because negotiations are repeated, and those that spring from the personal and professional codes of ethics that exist in the mediation field. In international relations, diplomatic training and concerns for reputation support the norms of conduct for mediators. Once again, mediation in environmental-development conflict lacks the support of either institutions or practices that can regulate mediator actions.

When negotiations are repeated, the bargainers can control or limit the use of discretion by third-party intervenors, and mediators can limit the incentives inducing bargainers to manipulate the mediator. Although either mediators or negotiators may successfully exploit the mediation relationship in a particular negotiation, the repetition of negotiations can protect the injured party. If, at the conclusion of bargaining, a negotiator either discovers or suspects collusion between his or her opponent and a mediator, the negotiator can attempt to redress the inequities of the settlement in future bargaining rounds and reject intervention by outside parties. The fact that collusion between a negotiator and mediator can both poison the future bargaining relationship and eliminate the potential for the future use of mediators deters exploitation of this relationship for potential short-term gains. Similarly, mediators who feel that a negotiator has attempted to manipulate their actions may refuse to continue their intervention or deny their services in future bargaining. Thus, both negotiators and mediators can use repeating negotiations to insure that an individual bargainer or mediator cannot readily exploit the confidential character of mediation to advance the interests of one side in a dispute (Simkin, 1971).

Constraints on discretion can also arise from the internal codes of ethics that inform the actions of the mediators. A mediator is almost constantly faced with a dilemma, deciding when to exert influence over the course of negotiations, and when to act as a simple observer or monitor. Although this choice is inherently a judgment call, mediators can develop internal principles that help them to make decisions concerning these choices. Simkin (1971) argues that a mediator's primary commitment is to the negotiation process, that is, to the resolution of a dispute through the efforts of the negotiating parties. Simkin (1971) even says that in the selection of mediators, it is more important to choose individuals who support the problem-solving efforts of others than to choose those who like to solve problems by themselves. The objective of negotiations is to get the disputing parties to agree to a solution, rather than to have the mediator either arbitrate or judge the particular dispute. Thus, there is a shared view that mediators should play a passive role in developing the issues and compromises, but an active role in controlling the interaction of the disputing parties.

If one believes that there exists a contractual relationship between negotiators and the mediator legitimating his or her intervention, then a key factor in this relationship must be a shared perception of what types of mediator action are appropriate. Negotiators often use their own discretion in determining whether to accept mediation services. If this decision is made with an understanding of what types of actions a mediator will take, then mediators who limit their actions to these expected interventions will have an implicit legitimation of their use of discretion. Unfortunately, when negotiations and mediation take place for only one time, as in many development disputes, there is little opportunity to develop a shared perception of appropriate mediator interventions.

Internal checks on a mediator's use of discretion arise not only from a reasoned analysis of what constitutes appropriate action, but also from an individual's innate understanding of what one should do in a particular setting. In our dealings with people, we find that there are some individuals who always seem to do the right thing while others seem doomed to perpetual error. Likewise, the personalities of some individuals prove especially suited for the task of mediation. In particular, some individuals will inspire trust among the negotiators that will open communications channels. Other individuals will raise suspicions of the negotiators and actually become an impediment to a settlement. An action may cause bargainers to feel that one mediator is biased or incompetent, yet they may dismiss the same action by another mediator as a simple mistake without serious consequences for negotiation. Thus, the bargainers' perceptions of the mediator can have a critical impact on the success of negotiations.

Although it is natural to focus on the personality of a mediator, care must be taken to avoid underestimating the importance of skill and experience. Simkin (1971), after a long career as a labor mediator, devotes only two pages in his book on mediation to personal qualities. Although the personal qualities of mediators are important for negotiations, one should not believe that personal character traits can be a good substitute for training, skill, and experience.

Various writers on mediation have described the personal attributes that may prove an asset to a mediator. Deutsch (1973), a psychologist writing on mediation and third-party intervention in disputes, states: "third parties can help in resolving disputes constructively to the extent that they are known, readily accessible, prestigious, skilled, impartial and discreet" (p. 388). He notes that these notions are ambiguous, particularly the notion of *skilled*. He suggests that this ambiguity arises, at least in part, from the difficulty in defining the criteria for successful mediation.

Landsberger (1955, 1956) and Simkin (1971), in their studies of the mediation of collective bargaining, have formulated lists of mediator qualities. Landsberger (1955), in his studies of labor mediation, has found that there is a large amount of agreement among collective bargaining participants in assessing the skills of a mediator. Through interviews, he built a set of criteria which participants in negotiations commonly used in assessing the performance of mediators. These ten areas are:

[1.] originality of ideas;
[2.] a sense of appropriate humor;
[3.] ability to act unobtrusively;
[4.] the mediator as one of us;
[5.] the mediator as a respected authority;
[6.] willingness to be a vigorous salesman when the situation requires it;
[7.] control over feelings;
[8.] attitudes toward, and persistence and patient effort invested in the work of mediation;
[9.] ability to understand quickly the complexities of a dispute; and
[10.] accumulated knowledge of labor relations. (p. 554)

Simkin (1971), in a lighter vein, listed the following mediator qualities:

1. the patience of Job,
2. the sincerity and bulldog characteristics of the English,
3. the wit of the Irish,
4. the physical endurance of the marathon runner,
5. the broken-field dodging abilities of a halfback,
6. the guile of Machiavelli,
7. the personality-probing skills of a good psychiatrist,
8. the confidence-retaining characteristic of a mute,
9. the hide of a rhinoceros,
10. the wisdom of Solomon.

And more seriously:

11. demonstrated integrity and impartiality,
12. basic knowledge and belief in the collective bargaining process,
13. firm faith in voluntarism in contrast to dictation,
14. fundamental belief in human values and potentials, tempered by the ability to assess personal weaknesses as well as strengths,
15. hard-nosed ability to analyze what is available in contrast to what may be desirable,
16. sufficient personal drive and ego, qualified by willingness to be self-effacing. (p. 53)

It is important to note how the qualities mentioned complement the basic sources of successful mediator actions. The originality of ideas, the ability to understand issues, vigorous sales ability, and accumulated knowledge enhance a mediator's *negotiating skills*. His or her ability to act unobtrusively, confidence-retaining abilities, thick hide, and demonstrated integrity and impartiality are qualities that support his or her *communications activities*. The authority of a mediator enables him or her to use the *power of position* to support agreements.

Controlling Discretion in Environmental Negotiations

Although it is easy to list examples of inappropriate actions in a negotiation over a development dispute, it is much more difficult to develop criteria that one can use to guide the conduct of negotiators and mediators. No traditions, such as those in labor or international relations, exist to guide the actions of negotiators. No laws clearly set the limits of appropriate action and tactics. Rarely do existing relations permit mutual expectations to serve as a guide to appropriate conduct. More rarely do either formal or informal sanctions discourage abuses of discretion or authority.

Several attempts have been made to address those special issues that arise from the exercise of discretion in environmental mediation and negotiation. The most ambitious efforts to analyze the problem of accountability in environmental mediation derive guidelines from underlying principles. Cormick (1982) starts from the principle that "the basic ethical principle that the intervenor should espouse is self-determination for all parties to a conflict" (p. 1). Cormick then argues that three elements critical to self-determination are (1) *providing information* concerning the intervenor's approach, the process itself, and a means for assessing whether to use it; (2) insuring that participants will have the *power* to bargain as coequals; and (3) providing the opportunity for *involvement* by all affected parties in a dispute. He argues that unlike in labor mediation, one cannot rely on the ability of the bargainers to enforce norms of conduct, and that mediators in a development dispute bear a greater responsibility for their conduct. He then proceeds to develop a series of questions to guide a mediator's actions.

Susskind (1981) holds a more ambitious view of the potential for mediation in resolving disputes and therefore sees the professionalism of environmental mediators as an inadequate safeguard for insuring their accountability. He uses Fisher's notion (1979) of principled negotiation to develop three guidelines for mediators:

> Working back from these [Fisher's] measures of success, it is clear that environmental mediators ought to accept responsibility for ensuring (1) that the interests of parties not directly involved in negotiations but with a stake in

> the outcome are adequately represented and protected; (2) that agreements
> are fair and as stable as possible; and (3) that agreements reached are inter-
> preted as intended by the community-at-large and set constructive prece-
> dents. (Susskind, 1981, p. 18)

Susskind believes mediators bear a responsibility both to the negotiators
and to society as a whole. His article contends that the mediator is the
key figure for guaranteeing the legitimacy of the negotiation process and
protecting larger societal interests. Susskind argues that given the large
scope of a mediator's duties, accrediting mediators (a technique recom-
mended by Cormick) may prove an inadequate method for insuring that
a mediator in an environmental dispute will act both effectively and eth-
ically. He argues that a more appropriate way to control mediator action
is to link mediators with regulatory agencies and the courts, and to create
an informed public that can immediately scrutinize a mediator's actions.
Susskind calls for legislation that could describe the circumstances and
conditions under which mediation could take place, as well as guidelines
for mediator actions. He sees government's role as one of insuring the
legitimacy of the negotiation effort and of protecting larger public interests that may go unrepresented at the negotiation table. Courts and
administrative agencies would house, supervise, and finance mediation
efforts. Susskind argues that as a long-run strategy:

> The most effective way to hold environmental mediators accountable would
> be to increase the public's capacity to demand fair and effective behavior on
> the part of mediators. (p. 44)

One's approach to the issues raised by the discretionary power of
mediators and negotiators depends on the scope of that one envisions for
this method of resolving disputes. If it remains as an ad hoc alternative
to litigation that is used only under favorable conditions, then perhaps
the training and accrediting of mediators may prove adequate for controlling the uses of discretionary authority in negotiations. However, if
one sees negotiation in development conflicts as an alternative whose
potential approaches that of negotiation in industrial relations, then the
more ambitious approach of Susskind is warranted. Accrediting alone is
unlikely to provide either the supervisory control over mediators or the
consistent concern for larger public interests that the more direct use of
government authority will likely produce.

In general, there are several relationships in negotiation in which
the exercise of discretion can raise questions concerning the accountability of the bargaining process and the individuals involved. These
include:

1.   The relationship of the negotiation process and its outcomes with the external community not directly represented in the bargaining;
2.   The relationship of a bargaining representative with his or her constituency;
3.   The relationship of the negotiators with each other;
4.   The relationship of the negotiators with those not represented at the bargaining table;
5.   The relationship of the mediator to the negotiators; and
6.   The relationship of the mediator with the larger community.

Both Cormick and Susskind focus on the relationship of the mediator to the negotiators, and both raise issues concerning the relationship of the negotiations to larger societal interests. Although not explicitly incorporating this six-point structure in his article, Susskind's suggestions for handling the problem that he identifies as "mediator accountability" include measures to limit the realm of negotiator and mediator discretion both by law and by systematic supervision of mediators employed by government. Susskind sees the mediator as having a special duty to protect the public interest in negotiation by attempting to insure that negotiated settlements will not contravene existing legal precedents and will not set inappropriate precedents for the future.

In my view, if one hopes to use mediation and negotiation more widely in development or planning disputes, then it becomes more important to legitimate and limit the exercise of discretion by both mediators and negotiators. The way to do this is with laws that create bounds for the exercise of discretion, rather than to single out any participant for either special control or special duties. Once again, labor–management laws and practices provide an example of how one can both permit discretion and limit it. Laws regulate union elections and strike practices and prohibit racketeering. These laws moderate the relationship of a union leader to membership and to bargaining opponents and they set limits on his or her actions. Similar provisions limit the actions of management officials. Mediator actions are controlled by codes of professional ethics, negotiator scrutiny, and the managerial actions of federal and state mediation services.

Attempts to resolve these issues of discretion should focus on holding the negotiation process accountable to the larger society and its concerns. Although the mediator's position will prove more important in development conflicts than in labor–management relations because of the one-shot nature of many of the conflicts, there is no particular reason to burden this one bargaining participant with managing the process on society's behalf. In the ad hoc negotiations now practiced, the participation of the mediator is often the critical factor in both starting and structuring negotiations. Thus, the mediator's action raises special ethical concerns. To sustain a wider role for negotiations, policymakers must reduce the dominance of this figure and enhance the role that negotiators play independently.

Conclusion

The exercise of discretion by negotiators and mediators can determine who benefits or loses from negotiation. The exercise of discretion raises several issues concerning the legitimacy of using negotiation for resolving disputes over projects that affect the welfare of many who do not bargain directly:

- Can one develop procedures that enable competing groups to recognize who has a legitimate role in a collective decision?
- Can one develop limited agendas that people accept as legitimate yet that are rich enough to encourage a cooperative resolution of a dispute?
- Can one develop principles that legitimate yet constrain the exercise of discretion by the representatives of bargaining groups and by mediators?

In collective bargaining, labor laws and tradition have combined to determine legitimate bargaining participants and to set bargaining agendas. Labor law acts to hold representatives of business and labor accountable to their constituencies. These laws limit the domain of negotiation discretion. Federal law establishes a federal mediation service that provides mediators to assist negotiators in resolving disputes. The repetitive character of labor negotiations and the professional norms of the mediation service provide checks on the inappropriate use of mediator discretion. In addition, the selection and training of mediators help them to develop qualities that serve as internal checks on the exercise of discretion. Similar actions hold the greatest promise for controlling negotiator and mediator discretion in negotiation over development disputes.

CHAPTER 9

Necessary Elements for Mediation

Introduction

In the best of all circumstances, negotiation and mediation can enable disputants to resolve disputes by finding consensus. At other times, only difficult compromises may make a negotiated settlement possible. In either case, a mediation process may offer a particularly attractive and productive alternative to destructively adversarial modes of decision-making. Mediation and negotiation, however, cannot resolve all disputes. In the worst cases, failure to successfully conclude a negotiation or mediation effort can actually complicate the conflict situation or lead to higher levels of destructive interaction. In addition, the disputants may blame the mediator for both the escalation of conflict and the failure of negotiations.

Since negotiations present risks as well as benefits, both potential negotiators and mediators must make critical assessments in determining whether or not to engage in a particular negotiation or mediation effort. Just as reaching a decision to accept a negotiated agreement requires individuals to assess the consequences of an agreement, deciding whether to negotiate at all requires a similar assessment. As long as adjudicatory modes of dispute resolution offer the accepted method of regulating environmental conflict, both competing groups and third-party mediators will need to determine whether the characteristics of a dispute warrant an attempt at resolution that relies primarily on negotiation.

All participants should assess how the various elements of a dispute affect negotiations. In particular, disputants should assess whether participation in a particular negotiation will advance the resolution of a conflict, add to existing animosity, or jeopardize their position in future litigation. Furthermore, the mediation effort itself may impose costs, such as time and money. Negotiations and mediation will require skills and talents quite different from those needed or available for litigation. Thus, organizations may need to develop skills quite different from those they normally possess, and the negotiation process itself can place severe stress on an organization.

Although this book has described many successful efforts to mediate the resolution of conflicts, one can perhaps learn more from the failures of mediation and negotiation than from the successes. In a successful negotiation, one sees the convergence of negotiation processes, issue structures, and bargaining skills to produce a climate in which a negotiated compromise eventually appears as a likely and even necessary outcome. In the Camp David negotiations discussed earlier, one can see that Carter, Sadat, and Begin skillfully transformed the issues of the conflict, deferred irreconcilable issues to future negotiation sessions, and skillfully used the United States to moderate discussion and to guarantee key elements of the settlement. This fortuitous convergence fails to suggest how important each element can prove to the success of negotiations, and how rare it is to find the necessary convergence of bargaining institutions, issues, and leaders in either international relations or environmental-development disputes.

An examination of failed negotiations can help one to assess when and whether to attempt to resolve a dispute through negotiation or mediation. International relations offers some of the more dramatic and visible examples of failed negotiation efforts. This chapter will consider the Falkland Islands war. One can here observe the failure of several mediation efforts to halt hostilities short of conflict. This analysis can enable us to draw lessons for the emerging field of environmental mediation.

Falkland Islands War and Mediation Efforts

On April 2, 1982, Argentina seized the Falklands (Malvinas) and South Georgia Island, located in the South Atlantic about 300 miles off its southern coast. These military actions ended efforts that had begun in 1966 to solve conflicting claims of sovereignty by Great Britain and Argentina through negotiations under UN auspices. A diplomatic dispute that had many elements of a farce—the islands had little economic value, and Great Britain sought a formula to gracefully end its control—quickly became a shooting war.

The Falkland Islands war had its immediate cause in a minor dispute over work permits that rapidly escalated in an environment dominated by nationalistic rhetoric and alarming rumors. On March 1, 1982, Argentina had warned that unless a speedy negotiated settlement was reached, Argentina "would put an end to negotiations and 'seek other means' to solve the dispute" (Keesing, 1982, p. 31526). On March 19, 60 Argentine scrap-metal workers, working on a contract from a British firm that was authorized by the British Foreign Office, landed on South Georgia to dismantle an old whaling station. After they raised an Argentine flag, a British Antarctic survey team requested that they leave and not return until they had received permission to enter from the governor of the Falklands, who was responsible for the administration of South Georgia Island. The survey team reported this event to British administration on the Falklands. On March 23, all but a dozen workers left the island.

From this point, diplomatic events rapidly escalated. The British government informed the Argentine government that the remaining workers had to leave or obtain permission to stay. Argentina claimed not to know that the workers had gone to South Georgia. Britain threatened the use of naval force if Argentina failed to comply. Argentina responded with a claim to the Falkland Islands and its dependencies. Ship movements were tracked by all, and there was speculation of a military confrontation. Great Britain moved its ship *Endurance* to South Georgia, and Argentina sent an armed survey ship to protect its workers.

On April 2, the worst fears of the international community were realized as Argentine troops invaded the Falkland and South Georgia Islands. Sixty British troops laid down their arms as part of a ceasefire negotiated in the face of overwhelming numerical superiority. The troops were flown to Uruguay, and eventually to Great Britain. By April 12, 10,000 Argentine troops occupied the Falklands.

Britain's response to these events was immediate. On April 5, a British fleet set sail on the 8,000-mile voyage to the Falkland Islands. Negotiations to secure the withdrawal of the Argentine forces began, but they failed to make progress. Britain increased its political and strategic pressure on Argentina. The 10 Common Market countries acted on April 10 to ban the importation of Argentine goods and to ban the export of weapons to Argentina. Argentina responded with a reciprocal ban. Britain announced on April 7 that it would declare a 200-mile maritime exclusion zone around the Falkland Islands. Argentina declared a special naval command and theater of operations around the disputed territories.

As the British task force of ships sailed south, the United States sought to help resolve the conflict before further escalation of violence. Alexander Haig, U.S. Secretary of State, led negotiations that proceeded with little effect. On April 25, the British forces recaptured South Georgia Island, with only one casualty—an Argentine sailor shot in a misunderstanding with a British soldier. On April 28, Britain expanded its exclusion zone around the islands to include aircraft and all ships. On April 30, Haig announced the collapse of the U.S. negotiation effort and blamed Argentina for intransigence.

Fighting escalated rapidly. On May 2, a British submarine torpedoed and sank Argentina's *Belgrano*; 386 Argentine sailors died. On May 4, an Argentine plane sank the British ship *Sheffield* with a single missile; 20 British sailors died. The sinking of the *Belgrano* destroyed a Peruvian effort to mediate the dispute. Simultaneously, a UN mediation effort picked up urgency but proved unproductive.

After the failure of these three mediation efforts, military struggle replaced all diplomacy. On May 21, British troops landed on the Falklands, and after heavy fighting, the Argentine forces surrendered on June 14. British losses in the military campaign totaled 237; Argentina lost 746.

These events raise serious questions for both international relations and mediation. How did events surrounding frigid islands in the South Atlantic escalate so rapidly into such an alarming and bitter conflict? Why did the mediation efforts of the United States, Peru, and the United Nations fail to halt this escalation? Why did events that indicated British military resolve and the potential high costs in life to both Britain and Argentina harden each side's resolve, rather than softening the bargaining demands?

Institutional Setting for Diplomacy

The ability of an individual or a party to resolve a dispute or act as a mediator will depend on the institutional setting that surrounds a conflict. In international affairs, the sovereignty of nations and their ability to use force limit the ability of international institutions to resolve disputes. There is no institution that can establish binding rules of law or norms for interaction. International institutions, however, can help disputants by offering a forum in which the nations can interact, by passing resolutions that formally urge a peaceful resolution, by providing mediators to help the disputing countries reach a settlement, and by helping the parties to implement or police the terms of any negotiated settlement. These actions, although not sufficient for avoiding destructive conflict, can assist those nations that seek to resolve differences without the use of force.

Argentina and Great Britain's dispute about sovereignty over the Falkland Islands simmered for years with little action. In 1947, following an Argentine naval expedition to the Antarctic which landed in the South Shetland Islands, located near the Falklands, Great Britain offered to submit the dispute to the International Court of Justice, but Argentina declined. When both sides agreed to negotiate their disagreements following a 1965 recommendation by the UN General Assembly, there were still available a wide variety of institutional forums to help resolve their differences. Through annual, direct face-to-face negotiations, Great Britain and Argentina attempted to establish a framework for governing the islands that would resolve the concerns of both sides. In these early negotiations, the major issues that emerged were Argentina's absolute claim to sovereignty over the islands and Great Britain's concern for the rights and liberties of the native population. The 1982 negotiations ended in a deadlock, with Argentina demanding monthly meetings and threatening to resolve the dispute through other means.

Argentina's initial invasion ended this negotiated search but started a more urgent search within international institutions to control this dispute short of armed conflict. This search enlisted the energies of the Secretary General of the United Nations, the Organization of American States, and third-party nations, especially the United States and Peru. Between each major battlefield action, a special diplomatic effort attempted to halt further escalation. With each failure of negotiations, diplomatic initiatives began with greater urgency in a new institutional setting, and with a new mediator.

Following the initial invasion by Argentina on April 1, 1983, the international dispute-resolution apparatus swung into action, performing perhaps better than in any recent conflict situation. The United Nations acted by passing Security Council Resolution Number 502 on April 3, 1982. This resolution, drafted by the British ambassador, stated that the Security Council of the United Nations:

1. Demands an immediate cessation of hostilities;
2. Demands an immediate withdrawal of all Argentine forces from the Falkland Islands (Islas Malvinas); and
3. Calls on the Governments of Argentina and the United Kingdom to seek a diplomatic solution to their differences and to respect fully the purposes and principles of the Charter of the United Nations.

Although a United Nations resolution carries no binding authority, such a resolution can help both sides to frame the issues in a dispute and can potentially serve as a basis for negotiations. This resolution, by demanding a cessation of hostilities, helped set an agenda for the negotiations that followed.

The United States and Peruvian Mediation Efforts

Immediately after the initial armed conflict, the United States offered its services to both Argentina and Great Britain to mediate the conflict. The U.S. State Department denounced the use of force and on April 5 offered its good offices to help mediate the dispute. The links between the United States and the two disputants were quite strong. In addition, the United States has played a major role in all hemispheric affairs since the presidency of Monroe. For these reasons, the United States possessed the trust of both sides that could serve a mediator well, as well as political principles that would legitimate its efforts to both international and domestic audiences.

As the British fleet set sail, President Reagan endorsed United Nations Resolution 502, urged a peaceful resolution of the dispute, and dispatched Secretary of State Alexander Haig to act as mediator. This choice of mediator signaled to all parties the importance of the dispute to the United States. Haig was the highest ranking official in the United States diplomatic corps. He had gained wide experience in foreign affairs as an assistant to former Secretary of State Henry Kissinger. Only the involvement of the president himself could have signaled a greater United States interest in the peaceful resolution of this dispute.

For the next several weeks, Alexander Haig virtually lived on airplanes as he shuttled over the 8,000 miles of Atlantic Ocean that separated Great Britain from Argentina: April 7—Buenos Aires; April 8—London; April 9—Buenos Aires; April 11-13—London; April 13—Washington; April 14—Buenos Aires.

Despite Haig's mediation efforts and the approaching deadline of the fleet's expected arrival, little progress was made. Argentina initially offered to withdraw its troops if Britain recalled its task force but insisted that its sovereignty over the islands was not negotiable. Margaret Thatcher, Prime Minister of Great Britain, insisted that the withdrawal of troops take place *before* halting the fleet. She refused to make any immediate concessions on the issue of sovereignty and claimed that the issue of sovereignty remained unaffected by Argentina's military actions. This position, of course, did not preclude a resolution that would eventually give Argentina sovereignty over the islands.

This initial exchange of positions left room for negotiations. The dispute over separating the armed forces had been transformed from a debate over who should withdraw to a discussion of the timing and sequence of reciprocal troop movements. Subsequent negotiations focused on mechanisms for supervising an orderly withdrawal of troops and halting the fleet. Buenos Aires newspapers reported that President Galtieri, the head of the Argentine junta, had made proposals to Great Britain on April 19 including "i) withdrawal of forces behind a 400 mile ring around the islands, ii) formation of dual British-Argentine administration, iii) replacement of Argentine troops by a jointly appointed police force, and iv) negotiations on sovereignty within a United Nations framework" (Keesing, 1982, p. 31535). Britain's Prime Minister Thatcher announced that she felt that these proposals failed to meet her concerns for the self-determination and self-rule of the Falklanders. Nevertheless, discussion between Great Britain and Argentina continued.

On April 23, Great Britain (but not Argentina) agreed in principle to a United States plan that called for

> A cessation of hostilities; withdrawal of both Argentine and British forces; termination of sanctions; establishment of a US-UK-Argentine interim authority to maintain the agreement; continuation of the traditional local administration with Argentine participation; procedures for encouraging cooperation in the development of the islands; and a framework for negotiating a final settlement, taking into account the interests of both sides and the interests of the inhabitants. (Keesing, 1982, p. 31709)

Argentina failed to respond to this offer, and on April 25, Great Britain recaptured South Georgia Island through a military campaign that led to the loss of one life. On April 28, Great Britain declared that on April 30 it would enforce a total exclusion zone around the Falkland Islands, thereby banning any ships or planes from this region. This announcement was logistically necessary for future military action and served to provide the negotiations with another deadline. On April 29, Argentina refused to accept Haig's offer and instead called for a further clarification of the United States proposal. On April 30, Haig blamed Argentina's failure to compromise for the collapse of the negotiations, announced the end of the American mediation effort, and offered United States support for Great Britain's effort.

From press accounts, it is unclear whether Argentina rejected the United States proposals because of British military actions to recapture South Georgia, whether Britain launched the military attack because of the failure of negotiations, or whether these military events proceeded independently of the negotiations. Hastings and Jenkins (1983) state that the decision-making process in Argentina had collapsed, and that Admiral Anaya, the leader of the Argentine Navy, had held out for battle. Although it is impossible to know the true sequence of events and decisions without public revelations from the participants, Haig's bitter denunciation of Argentina and the subsequent inability of its military government to take needed actions suggest that the negotiation failure preceded the attack on South Georgia Island. Hastings and Jenkins (1983) suggest that British politicians needed a military victory both to strengthen their diplomatic hand and to meet the expectations of British citizens for a use of force. The subsequent British record of military and diplomatic action also suggests that they were very skilled at linking diplomatic initiatives to military actions. The incremental escalation of the conflict signaled that Great Britain was able and willing to gain its goals through military means. In any event, the capture of South Georgia with almost no loss of life (one sailor) and the good treatment offered to Argentina's military officials by the British troops preserved the possibility that a negotiated settlement would still prove possible.

Unfortunately, Argentina failed to respond to this clear signal that Great Britain would use military force to advance its interests. On April 30, following the final collapse of Haig's mediation effort, the British began the bombing of airfields on the Falkland Islands. Peru immediately stepped forward, offering its services as a mediator, and attempted to fashion a nonviolent resolution of this dispute. Once again, an individual nation was acting to avoid the destructive conflict that loomed on the horizon.

On May 2, Peru proposed a settlement plan similar to Haig's proposal, and Argentina requested modifications. Peru redrafted its proposals to include provisions that called for a consideration of resolutions made by international organizations regarding the Falkland Islands. The principal intent of this demand was to take into account an April 28 Organization of American States resolution that recognized Argentina's sovereignty over the islands. In addition, Argentina demanded that the implementation of the agreement include various countries agreeable to both parties. On May 2, a British submarine sank the *Belgrano*, killing 368 Argentines. This action ended any hope for a settlement under the terms and proposals that Haig and the Peruvians had set forth.

On May 4, an Argentine plane sank the British frigate *Sheffield* with a single French Exocet missile. On May 5, the Thatcher government, shocked by the loss of the *Sheffield*, agreed to accept Peru's plan but expressed little interest in the then nascent UN mediation efforts (Hastings and Jenkins, 1983). On May 5, Argentina formally rejected the Peruvian proposals. Thus ended this brief, although intense, mediation effort.

The Peruvian mediation effort offered Argentina a second chance to resolve the conflict short of armed struggle. This mediation effort was extremely important since Haig's withdrawal and public denunciation of Argentina gave them a diplomatic shock that should have provoked a reassessment of their position. Argentine officials had expressed surprise at the U.S. action, and the Peruvian mediation offered them another opportunity to avoid continued armed conflict.

The United Nations Mediation Effort

The sinking of the Argentine boat *Belgrano*, followed closely by the sinking of the British frigate *Sheffield*, gave both countries as well as the world community an indication of this dispute's potential for destructiveness. After these bloody encounters, many anticipated that the next step for the British would be invading the islands and retaking them from the Argentine forces. Such a military action almost always produces high casualties. In this lull before the eventual escalation of the conflict, the UN attempt to mediate the dispute advanced. On May 5, as Argentine and British forces skirmished through a series of commando raids, Pérez de Cuéllar, Secretary-General of the United Nations, started his effort to mediate a solution to this conflict. He began private consultation with representatives of the United Kingdom and Argentina. On May 17, amidst a series of domestic charges of a sellout, Britain put forward another set of proposals for halting the immediate conflict without prejudging the final resolution of the dispute. These proposals were slightly tougher than the concessions offered immediately following the sinking of the *Sheffield*. Hastings and Jenkins trace this tougher stance to political pressure in the House of Commons from conservative members of Thatcher's party. Pérez de Cuéllar passed these new proposals on to Argentina. On May 19, Argentina rejected these terms, and on May 20, the Thatcher government withdrew the proposals while charging that Argentina had bargained with "obdurancy and delay, deception and bad faith." Peru on May 20 presented both sides with fresh proposals, but on May 21, 5,000 British troops landed in the Falklands. Fighting lasted until June 14, when the Argentine forces surrendered to the British at Port Stanley. During the course of the Falkland military actions, Britain lost 237 service personnel and 18 civilians. It is estimated that Argentina lost 746.

It is difficult to fault the international community for this deadly conflict. This dispute did not escalate to armed fighting through the inattention of the international community, or because of a lack of skilled mediators. Although international institutions have often failed to help disputants until a conflict has progressed too far, the United Nations, the United States, and Peru offered mediation services before major military action. Although the simultaneous efforts of Peru and the UN may have led to some initial confusion, the timing of the major thrust of each mediation effort allowed both parties to reassess their position in light of both the new military situation and prior negotiation failures. One must look elsewhere to explain the failure of negotiations and reason to prevent the escalation of a minor dispute into a shooting war.

The Falkland Islands: The Issues and Stakes

The way in which a conflict arises can affect the ability of the parties to resolve it through negotiation. This is as true in international disputes as it is in family fights. When a conflict arises over issues that are amenable to compromise, then it is often easy to achieve a negotiated settlement. When, however, conflicts take a polarized form that creates clear winners and losers, then no negotiated solution is likely to resolve the dispute.

Although the question of which nation would have sovereign authority over the Falkland and South Georgia Islands presents an issue that possesses an extremely polar form, the early negotiation efforts sought to overcome these problems by focusing on the mechanics of local governance, enumerating the rights that the Falkland Islands residents would possess under Argentine rule, and establishing a mail service between the islands and Argentina. Indeed, throughout the early negotiations and through those that took place after the seizure of the islands, many newspaper accounts called for a Hong Kong solution. Under this scenario, Britain would achieve through negotiations some formula for recognizing the claims of Argentina yet permit the populace to preserve their rights and liberties.

At its best, the control of the Falkland Islands presented Great Britain with a poor economic bargain. Although the wool produced by the islanders is among the best in the world, the great distance of these Islands from Great Britain made them quite expensive to manage. Although rumors of potential offshore oil abounded throughout the press, as far as I can determine there is no hard evidence of oil, and this did not seem to play a major role for either Argentina or Great Britain. The historical roots of the dispute very likely had more impact on the course of events than any modern concern for economics or oil. Britain had almost gone to war with Spain in the eighteenth century because of Spain's efforts to expel British settlers. Argentine children learned of

Great Britain's unjust seizure of the Falklands, and Perónists had periodically used this issue to whip citizens into a nationalistic fervor.

The invasion of the Falkland Islands by the Argentine forces caused the dispute to take a particularly concrete form. This action hardened the positions of both sides concerning their claims of sovereignty, gave the issues an importance that made compromise difficult, and made immediate action necessary. In addition, Argentina's action created a second set of issues over what constitutes an appropriate use of and response to force, and how to organize a withdrawal of troops from the islands without losing face. The use of force had made obfuscation of the underlying dispute impossible. In national politics, one cannot easily marshall forces for vague positions or diplomatic subtleties, nor abandon positions after the shedding of blood. Thus, as the conflict escalated, the use of force made negotiation more difficult.

Although the issues of this dispute complicated any efforts to achieve a negotiated settlement, by itself the dispute did not pose an insurmountable obstacle to its resolution. Imaginative or concerned leadership might have proved capable of establishing some form of trust arrangement or might have produced some form of guarantees that would overcome the hostility of the Falklanders to Argentine control. (However, little progress was achieved either in the UN negotiations prior to the dispute, or in the mediation efforts under the deadline of impending war.) Egypt and Israel had negotiated more difficult issues in the Middle East. In retrospect, the immense costs of the war to both sides, both in lives and in money, would have made almost any other resolution better for Argentina, and perhaps for Great Britain as well. Of course, no one would believe that Argentina would seize the islands, or that Great Britain would wage a war to recover them. Perhaps the failure of each side to seriously consider the demands and concerns of the opposing side led each side to fail to devote creative political and diplomatic energies to this dispute before it reached a military confrontation.

Failures of Leadership

In order to end a dispute through negotiation, the opposing parties must have a leadership that has the ability to define issues, develop alternatives, make concessions, and commit the nation to the terms of settlement. The Falkland Islands dispute became a shooting war because of failures of judgment and leadership. The Argentine military junta proved most incompetent in its decision-making capabilities. Of all the obstacles that worked to prevent a peaceful resolution of this dispute, perhaps the single most important was the failure of the Argentine leadership to make practical or informed decisions when they proved necessary.

The seizure of the Falkland Islands had a domestic as well as an international dimension. On March 30, three days before the invasion of the Falklands, mass demonstrations against the continued military leadership took place throughout Argentina. The day after the invasion of the islands, Argentina was once again full of demonstrators, only this time jubilantly announcing their support for the military junta's action against the Falklands (*Latin America Weekly Report*, April 9, 1982, p. 11).

The invasion not only served to quell domestic unrest but also solved a diplomatic problem that was just then coming to a head. Britain's protest to Argentina of the presence of construction workers on South Georgia Island presented Argentina with a difficult diplomatic problem. To either recall the workers or complete the paperwork would, in Argentina's view, constitute an implicit recognition of the British claim to sovereignty over these islands. Acting on this issue, Argentina seized both South Georgia and the Falkland Islands. This single military action solved both the internal domestic problem and the long-simmering international diplomatic problem.

The seizure of the island was a military solution by a military government to a political and diplomatic problem. It stemmed not from leadership strength and calculation, but rather from weakness and misperception. In retrospect, this proved not only a diplomatic error, but a military and domestic error. Not only did the Argentine generals miscalculate Great Britain's willingness to defend the Falklands, they grossly miscalculated their own ability to prevent recapture by the British and the domestic consequences of their own failures. The military defeat led to the replacement of the military leadership. This crisis, unfortunately, could have ended peacefully only if the government of Argentina had had the ability to exert strong, consistent, and skillful leadership.

Unfortunately for Argentina, its government failed miserably in its efforts to provide leadership either in negotiations or in military preparation. Argentina's government may be best viewed not as the dictatorship of an army general, but as the coalition of representatives of a semiautonomous army, navy, and air force (Hastings and Jenkins, 1983; Finer, 1975). This committee structure complicated Argentina's ability to negotiate and also injured its ability to stage concerted military actions that involved all three branches. The ability of these individuals to act was limited even further by the fact that members of the ruling junta had to report back to service councils. Each service conducted its own war and diplomatic effort against a concerted British military effort. The air force fought bravely, the navy failed to leave port, and the army's leadership proved unequal to the demands of battle. Each service locked itself behind a different political position. The air force favored compromise, the navy favored war, and the army had a center position. The diplomatic efforts of Costa Mendes, head of Argentina's foreign ministry, had an uncertain relationship with the junta.

Although an error of perception might have explained the initial seizure, it is difficult to understand why the Argentines seemed incapable of reacting to Great Britain's actions, whether diplomatic or military. There were several periods during which one might have expected the mediation efforts of third parties to succeed. The first opportunity came during the sailing of the British fleet. The fleet's imminent arrival at the Falklands offered a good deadline for negotiation and should have served to force a reassessment by Argentina of Great Britain's willingness to fight. Unfortunately, the Argentine regime did not act, apparently believing that the British forces would not fight once they arrived. This period was marked by major U.S. mediation efforts. The diplomatic initiatives failed to produce much progress, and Argentina's demands changed only slightly, calling for a withdrawal of forces and joint administration rather than its original demand for full control. Even the United States' eventual withdrawal from the talks and public support of Great Britain failed to introduce any radical reassessment of the situation. The generals did not want to admit their diplomatic or military miscalculation, and domestic propaganda generated a stream of statements regarding the situation that were false.

While the fleet sailed, no military action had taken place. Perhaps it is unrealistic to have expected a military government to have backed down from a show of force without some sort of fight. The quick seizure of South Georgia Island by the British should, however, have given the military government some idea of both Britain's military capabilities and its willingness to fight. The capture of the island with little loss of life offered Argentina and Great Britain a chance to stop the conflict without heavy bloodshed. Such action, however, would have posed domestic risks to the divided Argentine military leadership. The seizure of South Georgia was a military loss for the army, not for the more active proponents of the Falkland Islands war, reported to be the navy. Thus, the seizure of South Georgia produced little government action other than a series of press communiqués.

It is particularly difficult to understand the failure of the Argentine leadership to conduct negotiations following the initial sea and air engagements off the Falklands, but prior to the actual invasion of the islands by the British Marines. By that point, it should have been clear to all that the British were serious, and would go ahead with their planned actions. The sinking of the *Belgrano* both exacted a heavy cost in lives and caused the Argentine fleet to remain in port. Action at this time could have avoided a potentially bloody invasion by British forces, as well as the confrontation between British forces and Argentine forces, which were low on supplies. There was, however, little movement by the Argentine government away from their initial demands. They acted as if they lacked the ability to negotiate or bargain with the British, as if they were a committee of individuals with different stakes, simply lumped together but lacking the ability to negotiate or bargain with the British as a single unit. Rather, it appeared that the government was more concerned with appearances than the underlying realities, first unable to believe that the United States would ever support Britain, then unable to believe that the British would actually defend the sovereignty of the citizens on their islands, and still later unable to believe that individuals were fighting and dying because of the decisions (and lack of decisions) by their government in a time of crisis.

This failure of the military leadership suggests one of the fundamental problems of negotiations or mediation. Efforts to negotiate or mediate the resolution to a dispute cannot succeed if the leadership of a group are unable to make decisions or unable to change their positions to reflect new realities. In part, the three-service leadership of Argentina seemed to prevent the placing of responsibility for the overall war effort on one individual general. Coordination between services is always difficult, but the semiautonomous character of the Argentinian forces enabled the navy to withdraw after the sinking of the *Belgrano* (Hastings and Jenkins, 1983). As the war proceeded, each of Argentina's armed forces played a role independent of the realities of combat. The army invaded the islands and garnered the initial political and popular acclaim. This served to support the Argentine government, led by an army junta. The early capitulation of the army on South Georgia hurt the domestic power and reputation of the army. Only further military action could repair its reputation. The sinking of the *Belgrano* and subsequent failure of the Argentine navy to venture far from port seriously diminished the domestic power and reputation of the fleet, the strongest advocate of military action in the Falklands. The air force, although the smallest of the armed forces, gained a reputation for its heroic fighting. Each service leader appeared to think that the performance of his forces would redeem the reputation of the country and secure his own position.

There are some important lessons to be learned from this disastrous endeavor. Negotiation can succeed only if the individuals who lead groups have the capability of making decisions and acting on them. Although mediators can at times enhance the ability of leaders to make decisions and can at times demonstrate the underlying realities of force and politics to the negotiators, they cannot enable individuals and groups that lack the capacity to decide to become effective bargaining participants. Nothing that Alexander Haig did was able to alter the apparent vacuum of leadership at the top of the Argentine government. Even military events failed to cause a reassessment of the risks posed by the pursuit of a military strategy. The Argentine government marched fatalistically toward its doom, leaving the troops poorly supplied and hostage to its failure of leadership. The leadership appeared to believe its own propaganda, despite international reports that repeatedly debunked rumors and propaganda posing as truth. Hastings and Jenkins (1983) report that when the field general advised Galtieri that a battle could mean a total massacre, Galtieri ordered him to counterattack.

In our earlier discussion of bargaining leadership, we suggested that bargaining may prove most productive when the leadership of a group has the power and the ability to formulate bargaining positions, the ability to modify these positions, and the ability to balance the competing interests of subgroups within the larger group. Unfortunately for the Argentine state, it appeared that the leaders had initiated this foreign military campaign to distract the populace from its inability to undertake the balancing of interests and difficult decision-making needed to effectively run the economy of their nation. It is ironic that the war, rather than distracting the populace from the failure of leadership, served to make the failures more apparent. The reality of defeat exposed governmental lies and the sham of military power. A popular postwar slogan in Argentina was "los chicos murieron, los jefes los vendieron" ("the boys were killed, the chiefs sold them out") (*Latin America Weekly Report*, June 18, 1982, WR-82-24, p. 1).

In general, mediators will not succeed when the leadership of the conflicting parties lacks the ability to resolve the basic issues that constitute the heart of the conflict. Although mediation may serve to enhance the ability of leaders to initiate a negotiation, it will not enable or convince leaders with no bargaining ability or interest to do so.

Embarking on Negotiations, or Mediating a Dispute

The Falkland Islands war dramatically shows the failure of both negotiation and mediation to resolve a dispute short of armed conflict. This large failure raises many questions concerning the use of negotiation and mediation. In particular, disputants must determine whether and when to initiate negotiations, and mediators must determine whether and when to offer their mediation services in a particular dispute. In many respects, this analysis is quite straightforward. Disputants should examine the potential gains and losses in negotiation against the available nonnegotiated alternatives.

The potential gains and losses are often probabilistic rather than deterministic, and choices invariably pose risk. Once both Argentina and Great Britain had initiated military action, failure to negotiate the Falkland dispute offered the uncertainties of armed conflict. Armed conflict would likely produce high casualties and the destruction of expensive military equipment. A loss on the battlefield would likely end the political careers of the leaders of the losing side (as it did). Negotiations, however, could not restore the preinvasion status quo, and each side would likely need to alter its claims of absolute sovereignty over the islands. A priori, negotiations must have appeared better than the military alternatives, but there was no clear direction that negotiations could take, nor was it certain that Great Britain would undertake the risk of armed conflict. Both sides did embrace negotiations, but Argentina's military leadership clung to its initial demands despite military and diplomatic developments. This intransigence precluded the negotiators from determining a diplomatic course that would permit a constructive and practical resolution.

Mediators must also make strategic calculations before entering a dispute. Such calculations should include a consideration of the effect of their intervention on the course of the conflict, and of the effect of their participation on the stakes and interests of their own organization. In doing this, there are several questions to consider: Will their participation make things better or make things worse? Will mediation delay the negotiators from confronting the underlying realities that separate them and shape the overall dispute? Should the parties avoid an adversarial confrontation, or would an adversarial test of strength lead to a resolution of the dispute that would provide a basis for a lasting productive relationship? What effect will the success or failure of the mediation effort have? These factors should influence the decision of a mediator to enter a particular dispute.

In the Falkland Islands dispute, the mediation efforts of the United States, Peru, and the UN had both altruistic and practical rationales. A successful mediation effort would prevent the loss of life and limit the risks of wider escalation of the conflict. In addition, the armed conflict could serve only to divide the Western world. For the United States, an armed conflict would force a choice between Great Britain, a trusted

NATO ally, and Argentina, a state which had recently offered prospects for more favorable relations. More importantly, support for Britain risked a U.S. rupture with its Latin American neighbors, which almost unanimously supported Argentina's claim, and support for Argentina would both implicitly condone expansionist aggression by a military dictatorship and jeopardize U.S. relations with the Western European nations, which almost unanimously supported Great Britain. Thus, the United States had great interests and incentives to help mediate the dispute. Similarly, the mediation of the dispute by Peru or the United Nations would enhance their international prestige. Thus, the three major attempts to mediate the dispute met both altruistic and practical concerns of the United States, Peru, and the UN.

The timing of a mediator's entrance into a dispute or negotiations can affect the chances for reaching a settlement. Psychologists suggest that there are particular times in any dispute at which the positions of the parties freeze, and other times when the positions of the parties are fluid (Deutsch, 1973). Mediation can work best when the positions of the parties are fluid and the parties are willing to consider alternatives to destructive conflict resolution. This can occur at several times in the negotiation. In the early stages of a conflict, the positions of the competing parties are unlikely to be set in cement but have a natural fluidity as the issues are still emerging. Here, mediation can work to help the competing sides formulate their positions and concerns in ways that do not preclude negotiation. A mediator can help the parties to refrain from making public commitments to positions and may help the parties to avoid the polarization of the conflict through public statements.

If the first efforts at mediation fail, there is likely to be a polarization of the conflict that precludes a mediation or negotiation attempt. In the labor field, when it becomes apparent that a strike will take place and cannot be avoided, then the most effective thing for a mediator to do is to withdraw from a dispute and to wait until after the strike is mounted. Simkin (1971) says that when labor and management are clearly heading toward a strike, the leadership will be busy managing the strike preparation, making bargaining impossible.

When both parties reach a deadlock in conducting their dispute, then the new emerging realities can force a reassessment of their positions. As the bargaining positions begin to unfreeze, there exists a new opportunity for a successful mediation effort. Once a strike has been mounted and each side begins to face the harms caused by the failure to agree, there occurs another opportunity to negotiate as the positions and stands of the negotiators begin to unfreeze in the new reality that emerges from the costs imposed by the strike. A second mediation effort handled by another mediator can succeed where the initial effort failed. In the United States, this second effort is often managed by an individual of great political importance, such as the head of the mediation service, the secretary of labor, or the president. To date, environmental mediation has often taken place when the disputing parties have reached an impasse in their struggle. Only then does an attempt to negotiate a resolution begin to possess an attractiveness that warrants a good-faith bargaining effort.

In the Falkland Islands dispute, each mediation effort was initiated following a radical change in the relevant facts. The United States mediation effort began as the British fleet set sail for the Falklands. The Peruvian effort began following the U.S. denunciation of Argentina's failure to negotiate in good faith. The UN mediation effort followed the sinking of the *Belgrano* and the *Sheffield* and the collapse of the Peruvian effort. Each mediation effort offered the parties an opportunity to begin again with full knowledge of the new realities of the conflict. Despite timing that looked excellent from a theoretical view, Argentina's positions and views never had the fluidity necessary for productive bargaining. Even skilled timing of mediation efforts cannot overcome failures of leadership.

Conclusion

The examination of the issues, leaders, and mediation efforts in the Falkland Islands war holds lessons both for international disputes, and for disputes over development projects. In both environmental disputes and international relations, the institutional mechanisms that link the disputing groups do not encourage negotiation but permit and sometimes encourage destructive tests of strength. Although litigation is not as destructive as combat, the parties to an environmental dispute often lament the loss of time, resources, and energy absorbed in legal tests of strength. In retrospect, they often wish that they had attempted to resolve their differences through negotiation.

The issues in an environmental dispute often possess a polar structure that is commonly observed in international disputes. A stand for or against a particular development project resembles competing nations' claims to territorial sovereignty. Although different in content, opposition to proposed development projects and claims to the Falkland Islands both possess an all-or-nothing structure that can preclude any negotiated compromise.

In both arenas, claims are often tied to principles that can not be readily compromised. Environmentalists who link their positions to a principle of protecting sacred natural settings create disputes that are as difficult to resolve as those of the Argentines, who linked their Falkland war actions to historical claims to the Malvinas, and the British, who linked their actions to the rights of self-determination and self-governance of the islands' populace. Principles that are used to mobilize groups for special uses dramatically reduce the chances of reaching a compromise settlement.

Unlike the leadership of labor unions, neither the leadership of environmental groups nor the leadership of nations requires the leaders to possess the skills to represent a group in negotiations. When the leadership of either an environmental group or a nation views its task as providing an alternative vision of the world, then negotiation carries with it risks of corrupting this particular vision. For Argentine leaders, the invasion of the Falkland Islands conjured up both images of military power and their nationalistic vision of Argentina as a strong nation that obtains its objectives with strength of arms. The same invasion reminded British leaders of the failure of Chamberlain to oppose Hitler's aggressions. British democracy, however, served to ground the leaders' actions and fears in the realities that developed. In my view, they actively pursued a negotiated resolution of the dispute. Unfortunately for Argentina, its leadership appears to have believed its propaganda, feared its own people, and refused to consider unpleasant facts.

Successful negotiation requires a leadership that is capable of articulating its views, compromising its demands, delivering on its promises, and organizing actions to implement agreements. This necessitates making rapid judgments and calculations. Unlike in labor unions, it is far from necessary for the leaders of environmental groups or nations to possess these qualities. In both political and environmental arenas, willingness to compromise carries risks to those who rely on shared visions to maintain their leadership or group.

An examination of how one can decide to negotiate an international dispute can provide guidance for those determining whether to pursue bargaining as a means of resolving an environmental dispute. In both situations, bargainers must compare the costs and benefits of negotiation against alternative means of dispute resolution. Such a comparison requires that a group's leader assess his or her own bargaining skills and the group's ability to withstand the stresses of bargaining. Since the outcome of negotiations is always uncertain and because the failure to bargain often brings an uncertain result, this decision invariably requires that the leader compare alternatives that are uncertain and risky. Negotiations do not always succeed, and the effort to negotiate can affect resources available for future tests of strength. Labor mediators implicitly accept this fact and sometimes state that the relationship between the parties may require a strike from time to time. For them, avoiding a strike at all costs can actually limit or destroy the ongoing relationship. Similarly, in a dispute over development projects, at times the involvement of a mediator may actually delay each party in coming to the realization that they may need to work with each other in resolving the issues that link them. This can be particularly true when parties have adopted positions and strategies that aim not at resolving a particular dispute, but rather at altering long-term balances of power between the parties.

Finally, even when the chances of resolving a dispute through negotiation or mediation appear small, one must consider the costs of failing to try. In the Falkland Islands dispute, the failure to bargain led to a destructive war between Argentina and Great Britain. Although the mediation efforts of the United States, Peru, and the UN had slim chances of succeeding, almost any reasonable person would agree that these efforts were well justified. The success of one would have improved the lives of many. Their failure seemed to have no effect on the course of military events.

In environmental disputes, efforts to mediate a resolution will invariably face obstacles that lessen the chances of resolving a dispute. For many disputes, the high cost of court actions, delay, and uncertainty may warrant attempts to negotiate a settlement. When the costs of litigation are extremely high for both parties, then even negotiations that have a slim chance of success may offer an attractive option.

CHAPTER 10

Using Negotiation and Mediation

Introduction

This book has examined the uses of negotiation and explored its potential as an alternative to adjudicatory proceedings for resolving disputes over development projects. Currently, ad-hoc negotiation with mediation offers an alternative that has successfully supplemented traditional administrative and bureaucratic forms of dispute resolution in disputes over development projects. The successes of these initial efforts and the enthusiastic reception of this technique by dispute participants suggest that negotiation has a good potential in these disputes.

A comparison of the emerging structures of environmental negotiation with those of industrial and international relations has produced some disturbing conclusions. Despite the fact that bargaining takes place between citizens in the same country, the ad hoc methods of negotiating resolutions to development disputes share much with international relations:

- Disputing parties need not recognize opposing groups as having a legitimate role in the conflict.
- Groups need not negotiate but can readily resort to destructive contests. Negotiating agendas are difficult to set.
- Leaders are not chosen to negotiate and often lack the needed skills either to articulate interests or to compromise positions.

- The infrequent contacts between disputants can easily create a climate dominated by suspicion and distrust and can lead to a collapse of communications.
- Power may prove highly asymmetric.
- Deadlines seldom exist to structure bargaining.
- Mediators are available only on an ad hoc basis.
- Designing binding agreements can prove difficult.

This contrasts greatly with the bargaining environment of United States industrial relations, where:

- Law establishes rules for recognizing representative groups as bargaining participants.
- Law requires good-faith bargaining, restricts adversarial forms of conflict, and sets the bargaining agenda.
- Leaders are elected to represent unions in bargaining, and law establishes procedures to hold leaders accountable to members.
- Law establishes an ongoing bargaining relation, and work requires daily contact between labor and management.
- Contracts create bargaining deadlines.
- Law establishes a mediation service and aids the entrance of mediators into a dispute.
- Work contracts legally bind both parties.

This bargaining environment now aids labor and management to negotiate settlements and to manage conflict.

Policymakers face a choice that will determine the potential of negotiations for resolving disputes over development projects. Without the support of law or institutions, the use of negotiations in development conflicts will resemble international bargaining. Not only will the dispute environment provide little support for bargaining, but the successful uses of negotiation will remain rare events. Only by the selective targeting of negotiation and mediation efforts will bargaining develop a record of success. As in international relations, the lack of institutional mechanisms to control conflict makes bargaining especially dependent on the caliber of leadership.

The prospects for developing a leadership capable of bargaining are slim. As long as litigation offers an acceptable, trouble-free alternative to negotiation, then opposition groups and government agencies will seldom accept the heavy and often burdensome costs of training negotiators and undertaking the internal bargaining that successful negotiations require. In addition, without the help of law to set an agenda, efforts either to expand the range of negotiation issues or to negotiate compensation raise serious questions concerning their legitimacy. This can leave the leaders of negotiation efforts open to charges of selling out critical group interests, or of making extortionate demands that are outside the range of law.

A negotiation-based review process does, however, offer several advantages to those seeking to resolve development controversies. Negotiations can enhance communications between disputing interests. Discussion sessions can help each side to eliminate caricatures that often develop in a long dispute. The information provided by increased communications can enable community groups and developers to create agreements that leave them both better off than the resolutions reached by adversarial processes. Even when no agreement is possible, the good-faith discussions can help eliminate alternatives clearly inferior to available options. This may leave a better set of options for courts and administrative agencies.

Perhaps most importantly, a review process based on negotiations allows the direct participation of citizens in development decisions in a constructive way. Unlike traditional review proceedings, technical discussions of design features and legal precedents need not supplant the consideration of real interests or place the issues in the hands of lawyers and technicians. The communications involved in bargaining may also facilitate the direct compensation of individuals injured by development projects in the forms that they desire. When done within a framework endorsed by a political process, this compensation can enable the formation of a political consensus supporting or tolerating a needed development project. Such a consensus may reduce the political and regulatory uncertainties that a developer faces and may affect the probability of a later reversal of project approvals in the courts. Even when no consensus is reached, the process of negotiation may help clarify issues and alternatives better than current procedures.

Gains from negotiations and communications are, however, uncertain. Although negotiations will permit both direct communications between disputing parties and direct citizen participation in a development decision, direct communications may have disadvantages. The disputing parties could use direct communications to intensify their animosities. Even with the skillful intervention of a mediator, the intensity of a conflict may cause an explosion of passions and emotions that may eliminate the prospects of a constructive resolution of a dispute. An obstinate developer or community group may subvert any opportunity for resolving differences. Without cooperation between the competing interests, the interactions of negotiations may become adversarial.

Policymakers can design institutions and procedures to limit this destructive potential. A formal process of negotiations can both foster constructive interaction and limit destructive tactics. With the assistance of law, the negotiation of disputes over development projects can follow the example of industrial rather than international relations.

The Design of a Formal Negotiation Process

Our study of negotiations and the analysis of mediation identified several design issues that a formal negotiation process for resolving disputes over development projects must resolve if it is to consistently encourage constructive bargaining. These issues include:

1. What type of development decisions should be resolved via a negotiation-based review process?
2. What issues should constitute the bargaining agenda?
3. Who should participate in the negotiations?
4. Who should resolve procedural questions and chair the meetings?
5. How can negotiations encourage progress and prevent delays?
6. What determines the power of the negotiation participants? How should it be distributed?
7. How can agreements be officially recognized and disputes formally concluded?
8. How can one incorporate expert knowledge into a negotiation-review process?

The resolution of these design issues will have implications for the functioning of the negotiation process. I will discuss each in turn.

Which Development Decisions Should Be Resolved by Negotiation?

The first task of a policymaker is to determine which of the many disparate siting disputes should be candidates for formal negotiations. This question poses both practical and political issues. The acceptance of a negotiation review process as fair and efficient requires that those designing and implementing such a process give special attention to the development decisions that will fall under its review. Few responsible policymakers would desire to scrap existing procedures to try this novel approach in all new siting decisions. The risk of a large failure argues for a more incremental approach. Furthermore, out of fairness to both developers and communities, those designing a review process should also consider the rationale for government and public involvement in the review of a particular proposal. The appropriateness of using a negotiation process will vary with the characteristics of the developer, the economic environment within which a facility operates, and impacts of a facility's operations. Developers can include government agencies, regulated public utilities, and private firms. Projects will also differ in the size of their impacts, whether good or bad, on a community.

Public involvement in government project decisions has been long accepted. Government-sponsored projects require the expenditure of public funds raised by taxes. Further, these government facilities often impose costs or distribute benefits to the residents of the localities and regions in which they are sited. These considerations make government development projects particularly appropriate candidates for the public scrutiny generated by a negotiation process. Through a negotiation process, individual citizens may better express their views concerning aspects of the development decision and may use their power to affect both the decision and the design.

The development decisions of public utility industries, such as electric or telephone companies, offer a second class of appropriate candidates for negotiation. These companies have long been scrutinized by government and the public. Public-utility-rate commissions determine the structure and level of prices, as well as the need for new facilities and their location. Many states already require public hearings before revising rates or assenting to the construction of new facilities. A power plant review provides a dramatic example of the involvement of all levels of government in the regulatory process. Federal regulators review the plants to insure that their design will meet federal environmental and safety standards. State rate commissions determine whether or not local electric demand warrants the construction of a new facility. Local communities must provide key construction permits and zoning variances for the many incidental activities associated with the construction of the facility. Because of this already heavy government involvement, a negotiation process may offer an appropriate mechanism for resolving conflicts that arise over the siting and construction of energy facilities, both between the different levels of government control, and between the public and a particular firm.

All projects, no matter who develops them, will differ in the severity of their impact on a community. Negotiations, with their requirement of the participation of skilled individuals, may prove an inappropriate way of reviewing those projects that produce small impacts on surrounding communities. This may be especially true when private firms serve as developers. The government's role in reviewing such development projects is rarely large and seldom controversial. Although some citizens may prove deeply concerned with the design of a phone booth or a curb cut, preexisting procedures usually resolve these disputes without great controversy or uncertainty. Government review often involves routine actions to insure that buildings will be safe and that proposed projects will comply with local zoning laws. The replacement of simple procedures with a process that encourages elaborate public scrutiny of these decisions may be highly inappropriate. The use of a compulsory negotiation-review process to consider such development decisions may add needless costs to both government and developers. Thus a concern for developing an efficient procedure may caution against extensive use of bargaining for resolving disputes over small facilities that produce only minimal impacts.

The use of negotiations to review large projects that produce major local impacts may prove appropriate, even when the developer is a private firm. In addition to zoning laws, the statutory provisions of state and federal environmental laws have created a role for government in reviewing the economic and social impacts of local development projects. Oil and chemical plants, steel mills, and textile plants all face environmental and health reviews. The use of negotiation to consider the siting of private-sector facilities that produce great local impacts offers an appropriate way of addressing the public impacts of private development decisions.

No prudent policymaker would urge the wholesale adoption of negotiation to replace current review processes. In particular, a sequential implementation of a negotiation system may generate information concerning both its costs and its effectiveness that will allow administrators and legislators to assess the impacts of this alternative review process. This procedure can enable them to avoid disasters and to learn from inevitable initial mistakes. Because of the clear case for public involvement in government-sponsored development projects, these projects may offer the most appropriate initial candidates for review. For large-scale government projects, the costs associated with a negotiated planning process will constitute only a small percentage of overall project costs, and small improvements in the project or in the certainty of the siting decision can produce major benefits.

The largest single cost of negotiations may be the time that the developer, state officials, mediator, and voluntary participants spend in bargaining. The analytic and design resources devoted to the negotiations, as well as the administrative and bureaucratic resources used in formulating and certifying agreements, add to the costs of bargaining. These expenditures may prove the most appropriate when the contro-

Table 5
Appropriateness of Negotiations for Siting Decisions

Developer	Scale of impact	
	Large	Small
Government	Very appropriate	Less appropriate
Regulated utilities	Very appropriate	Not appropriate
Private firms	Appropriate	Not appropriate

versial nature of a project necessitates a close scrutiny of the project by the public as well as trained technicians.

Table 5 integrates these concerns for the nature of the developer and the extent of a project's impacts into a table that describes the appropriateness of using negotiations to review a particular project. Government projects producing substantial local impacts may prove most appropriate candidates for these reviews. Projects that produce great impacts on surrounding communities, no matter who proposes them, will also prove likely candidates for this approach. Small-scale projects producing few impacts on a community are probably best handled through the traditional zoning processes.

What Issues Should Constitute a Bargaining Agenda?

In establishing a formal process for resolving disputes over development projects, setting the agenda poses a question in judgment that requires an answer in moderation. If too few items beyond the most prominent issue are included in the bargaining agenda, then the disputing parties will be forced into a divisive form of zero-sum bargaining. If too many issues are enumerated for bargaining, then the task of finding a settlement may prove difficult to manage even with the best intentions. If statutes fail to clearly delineate a bargaining agenda, then questions will constantly arise concerning whether it is legitimate for the discussion to focus on a particular issue. The inability to clearly answer charges that an issue should not be discussed can eventually undermine public acceptance of negotiation efforts.

How does one determine which and how many issues should go into a particular negotiation? There is little empirical evidence concerning how the number of issues affects the likelihood of reaching a settlement. Bargaining theory, however, suggests that with a large number of issues, the opportunities of exploiting differences in preferences to realize mutual gains improve. As mentioned earlier, observers state that when there are several issues in bargaining, the disputants often begin by discussing principles and only later move to the issues (Zartmann, 1977; Iklé, 1964). This discussion of underlying interests can help the disputants move to a discussion that permits a wide search for agreement. On the other hand, the narrow definition of a dispute and its issues appears to lead to the collapse of bargaining unless an easy resolution is immediately apparent, which suggests that those establishing a system of formal negotiations should seek to make the bargaining agenda relatively rich. Thus, in a particular dispute, the agenda might best include not just the immediate effects of a project, but a wider discussion of socioeconomic impacts, including jobs created, local tax benefits, adverse effects on property values, and other local impacts.

The more technical issues, such as health and safety, pose particular difficulties for those setting a bargaining agenda. It is not realistic to expect to exclude them from negotiations. Nevertheless, the choices and risks posed by many projects often require a knowledge of technical details that are difficult for untrained individuals to grasp. However, not all safety issues require a mastery of technical details. Some safety issues will arise over a plant's operating procedures (such as the choice of trucking routes to a plant), and local residents may possess concrete knowledge that would escape the review of an abstract study. The provision of technical assistance to bargaining participants may aid the search for agreement.

Who Participates?

Those instituting a formal negotiation-review process must determine just who will participate in the bargaining. Meeting a goal of designing a fair siting process requires that all who are greatly affected by the development decision have the opportunity to participate in the negotiations. Existing procedures already establish a minimim level public participation that future review processes must meet. Nevertheless, determining just who participates, how diverse interests are consolidated, and how leaders are held accountable to a group's members leaves major issues for policymakers to resolve.

Several considerations support attempts to limit the number of negotiation participants. Large numbers of bargainers may make the negotiation process unwieldy and difficult to manage; negotiation sessions are unlikely to accomplish much when the number of bargainers is large. Additionally, in negotiations over environmental-development conflicts, many people will participate voluntarily. When the number of participants is large, the bargainers may feel that they will have only a limited effect on the final outcome.

In bargaining over development issues, many of the benefits of negotiation may arise only from an atmosphere of trust and understanding that develops through personal contacts between the disputants. Trust will prove more difficult to develop in a large group. Further, if participants can easily join and withdraw from negotiations, a climate of trust is unlikely to develop. People cannot constantly adjust to new faces.

The *developer's* participation in the bargaining is essential. The developer will make the final decision on whether to build a project with the mitigation and compensation desired by the local community, whether to abandon the project, or whether to develop in an alternative town. Thus, without a developer's participation, negotiated development plans would likely produce a series of formal guidelines that would fail to address the specific impacts of a real project. Participation in negotiations may enable the developer to gauge community sentiment and to assess the likely outcome of the postbargaining events. The developer's participation may enable the bargaining to depart from abstractions and focus on how the project will affect specific concerns of the community. The developer has access to vast amounts of information concerning the design of the plant and the costs of incorporating changes. The developer's participation in negotiations and his or her particular focus may allow the other participants to see the many problems that the developer of a large project must resolve. The access to design alternatives and the practical perspective brought by developers to a discussion may also promote constructive forms of interaction.

The inclusion of *representatives of government agencies* in negotiations may prove critical to the success or failure of this review process. Government agencies often possess veto power over the final project design or location. Without the participation of key government agencies, fear that change will jeopardize regulatory approvals may halt bargaining. In addition, postbargaining actions of nonparticipating regulatory agencies may unravel delicately balanced agreements forged by the negotiating parties. The participation of regulatory agencies in the evolution of an agreement may help avoid last-minute complications that may arise if changes are required. Even when the regulatory agency reserves its final decision for a formal postbargaining hearing, its participation can signal to the bargainers when proposals are exceeding the bounds of acceptability.

The participation of government agencies may also prevent the inadvertent disregard of some public interest not represented at the negotiating table. Many administrative review processes were designed to force some consideration of the public interest when developers make decisions. Although there is no single conception of the public interest, often a broad consensus exists over some particular aspects of a development project. In the siting of coal-fired power plants, for example, there is seldom controversy over whether pollution control equipment should be used, but there are often technical disagreements over whether standard pollution-control devices will enable a facility to meet standards that protect local air quality. These narrow technical questions of how to meet broad public concerns may best receive attention through the participation of public agencies with specialized skills. Furthermore, protecting public interests may produce negotiated settlements that require government agencies to implement some terms of the agreement. In these situations, the government agency acts as a major stakeholder in bargaining and its participation is central.

Selecting negotiation participants from *nonformal groups* presents a difficult problem in instituting a review system. Negotiation provides a major opportunity for public participation in the review of a project, yet which is the public affected by a project, and how would one select representatives? Projects may generate particular interest among local community groups which share their neighborhood with a project, regional groups which may receive the benefits of a plant's services and bear the impacts of its operations, and interest groups that have a special concern over a particular technology, facility, or site.

Negotiations should include representatives of neighborhood or community groups. These groups will most directly bear the visual and noise impacts of a facility. The physical environment near a facility will often be the one most affected by a project. Even when a plant produces great property-tax advantages for the town, the houses closest to a facility may suffer a depreciation in their property values and great inconvenience. For these and other reasons, these groups may desire special representation in a bargained review process.

A problem arises over how to recognize nonformal groups and interested individuals as participants in the negotiations, and how to determine who legitimately represents these groups. Whenever the negotiated review limits the number of participants in bargaining, some process must determine which of the many groups may negotiate and who shall officially represent them. In small communities with cohesive groups, it may prove simple to designate by statute some particular group or officeholder to represent these local interests, or to appoint someone for this purpose. In larger communities, there may not be natural groupings that can adequately represent local interests. Then, some procedure for eliciting effective local participation is necessary.

The desire to encourage participation must be tempered by the realization that unlimited participation may create cumbersome and unproductive negotiating sessions. A concern for the fairness of the process requires that a mechanism for limiting participation avoid arbitrary decisions. To encourage public participation, the method used to screen or aggregate individuals into groups should not impose heavy burdens on those who wish to participate. Thus, a screening or aggregation mechanism must offer both a natural way of limiting participation and a simple way for groups to formally designate an individual to represent their interests. The use of a qualifying petition may offer one such way of restricting participation to those with shared interests in a particular dispute. Those heading a petition would represent those who signed it. A minimum number of signatures would qualify an individual for participation.

Those designing the structure of such a process will face a trade-off between negotiation advantages gained through the consolidation of interests and the barrier to participation which a high qualifying standard presents. A low qualifying standard will facilitate participation, but in the extreme, it may produce an unwieldy number of participants. A low standard will enable many groups to generate the needed number of qualifying signatures internally, thus reducing the need of groups to reach out to others.

Finally, the participation of a *mediator* can prove especially helpful in development disputes. As described earlier, a mediator can bring negotiation skills and insights that enable him or her to aid first-time bargaining. Thus, designers of a negotiation process should develop a procedure that encourages early mediator involvement in a dispute or bargaining effort.

Who Resolves Procedural Disputes?

Once parties to a negotiation are selected, they must adopt some procedure for organizing and structuring their interaction. Although laws can determine the participants and the legitimate bargaining agenda, the law lacks the flexibility to determine bargaining procedures. Since the disputing parties will likely distrust each other, it is critical that some third party have the respect of the conflicting parties and authority that enables him or her to act to resolve disputes.

A mediator may provide his greatest contribution to bargaining by establishing negotiation routines and procedures. Although law can reduce the importance of a mediator by determining who bargain and what they bargain about, the mediator plays a powerful role in moderating the interaction of the negotiators with each other (Wall, 1981) and in indicating society's concern for a constructive resolution of the dispute (Simkin, 1971). Unlike the other participants, the mediator has no personal interest to promote. The major concern of the mediator is to assist the parties in reaching a negotiated settlement. As seen earlier, the mediator's control of the communications between bargainers can help create a climate which makes productive negotiations possible.[1]

Simkin (1971) relates that labor mediators often face major problems when they attempt to assist labor and management in first-time bargaining. Negotiations often take place after a bitter recognition struggle in which the management and union organizers adopt polar positions. When the union wins an organizing election, the management and union must then bargain with each other. The long-run productivity of this new relationship requires that they move away from extreme positions and search for a ground of accommodation. The mediator, in assisting at these first-time negotiations, must not only familiarize participants with negotiation procedures, but also try to assist the disputants in recognizing their mutual interests.

In development conflicts, the role of the mediator in orienting the bargainers will be very important. The negotiating parties will be meeting with each other in a nonadversarial context for the first time. They will generally lack the familiarity with negotiations that is often a critical element in successful bargaining. The mediator can orient the bargainers and assist them in their efforts to acclimate to the negotiations. This orientation may have particular importance after the introduction of the negotiation-review process since most participants will be more familiar with adversarial forms of interaction than with cooperative ones.

The mediator should have the power to maintain order at meetings, recognize speakers whenever informal procedures fail to work, and adjourn meetings. The mediator should moderate the levels of hostility between the opposing parties both by establishing the ground rules for interaction and by controlling the frequency of meetings. At times, a mediator may decide to meet separately with opposing parties, or to place them in separate rooms. Such control of the tempo of the meetings

---

[1]See Wall (1981) for a comprehensive list of mediator actions.

and interactions may create a psychological climate which supports an amicable negotiated settlement (Deutsch, 1973).

A mediator can also control the sequence of the issues addressed. Our earlier analysis showed the importance of the sequence of bargaining issues. The choice of the sequence of bargaining can facilitate the exploration of mutually advantageous concessions. Further, the advance announcement of the issues that a meeting will address may allow those less interested either to avoid the meeting or to anticipate that they will have little to contribute. The prior announcement of a topic will help the mediator to keep discussion from getting sidetracked.

How Can Negotiations Encourage Progress and Prevent Delay?

Insuring that a negotiation process will resolve a development dispute offers an efficient alternative to current adversarial procedures and requires that designers of this process take steps to improve its operation. In addition, considerations of fairness support a timely decision process with procedures known in advance. Delays often impose asymmetrical costs on groups involved in a conflict. If participants view bargaining as a tactic for delay, then it will prove especially difficult to gain the good-faith acceptance of a bargaining effort.

Setting a deadline structure for negotiations may offer the most promising approach to providing a timely review process and signaling to participants that negotiations are not meant to offer opportunities for delay. Negotiations without a deadline, which are common in international relations, can drag on endlessly, and lack of bargaining progress can become still another issue in a dispute. Although setting a rigid deadline risks losing agreements that are almost complete, in general the deadline will spur bargaining progress. In those rare situations when a settlement is close, a short ad hoc extension may add the flexibility needed.

As the deadline nears, the negotiators compare the results of failing to resolve the remaining issues with both the benefits that an agreement will bring and the cost of continued conflict. If they have made substantial bargaining progress, this calculation will differ greatly from one that they would have made at the start. In particular, the gains from a negotiated settlement will be clearer because the issues settled will help shape a concrete view of a settlement's benefits. Thus, the deadline can force a new assessment of the results of nonagreement, which will generally prove more favorable to continued bargaining. When little negotiation progress has been made, this calculation will look little different than that at the start of bargaining. In these cases, the deadline will officially and mercifully conclude the negotiations.

In ad hoc bargaining, the participants, in prenegotiation conferences, can agree to a deadline that will conclude a negotiation effort. This may serve as a substitute for a deadline set by statute, but it will likely prove less effective than an externally set deadline. If the negotiators set the deadline, then they clearly have the power to change it. Further, it poses still another issue that those initiating bargaining must resolve.

What Determines a Negotiator's Power, and How Should Power Be Distributed?

In bargaining, one's power at the negotiation table depends critically on what happens if one fails to reach an agreement. Although bargaining skills, preparation, and strategy may enhance one's ability to win concessions at the bargaining table, the basic structure of bargaining power is determined by the alternative to a negotiated agreement. In labor–management relations, both sides suffer the costs of a strike, but market conditions, strike funds, or the size of inventories determines which party is better able to sustain a strike. In international relations, the incentive to bargain depends on the consequences of failure both to the nation and to the individuals most responsible for the conduct of bargaining. Currently, the failure to reach a negotiated settlement in a development dispute leads to litigation. Litigation is also the path taken when negotiations are not attempted. As discussed earlier, it provides very weak incentives to either a group or its leaders for reaching a negotiated settlement.

Fundamentally, the question of distributing power to disputants remains a political choice, best handled by legislatures that are designing a negotiation process. To the degree that existing procedures liberally grant site-veto powers to affected groups, a negotiation process will necessarily reduce the power of those whose interests are served by vetoes. In my view, the justifications for limiting veto powers are essentially twofold: (1) the exercise of a veto may serve to undercut the interests of the community at large; and (2) the negotiation process offers a fair deal in which veto power is exchanged for fuller access to a decision process. As is clear from the character of these two arguments, their force will depend both on the specifics of the particular class of development projects decided through negotiations, and on the structure of the negotiation process that substitutes for the administrative and litigation procedures. Although experts can assess the implications of various institutional arrangements for the power of each bargaining participant, experts cannot guide the decision concerning the appropriate amounts of power to vest in the disputing groups. The judgment of a democratic legislative process offers the only legitimate way of resolving these issues.

There are several ways to limit bargaining power. One is to send deadlocked negotiations to an arbitration procedure that can result in a formal decision (O'Hare, Bacow, and Sanderson, 1983). This would then link bargaining power to a legislatively mandated arbitration formula. Another technique is to let a local legislative body or electorate consider the developer's last offer. Such a procedure would then link the bargaining power of a group to its ability to convince the legislative body or the voters that they should agree with its assessment that the final offer is inadequate.

Finally, when the disputants do reach a resolution of a dispute, there should exist a ratification process that certifies the negotiated settlement. A formal ratification of the negotiated settlement should help meet two objectives. It can provide a process that legitimates the negotiation by subjecting the agreement to wider public scrutiny, whether through a legislative review or a local referendum. In addition, on a practical level, unless there is some formal document that concludes the negotiation, the status of the final agreement will remain subject to uncertainty.

How Can Agreements Be Officially Recognized and Disputes Formally Concluded?

If the bargainers reach a settlement, they will need to incorporate the terms of agreement in a formal document. In labor–management collective bargaining, the contract contains all negotiated provisions, and the courts can enforce its terms. Treaties serve a similar purpose in international relations, but the lack of a higher authority makes the terms difficult to enforce. In ad hoc negotiations, the bargainers often have difficulty establishing a formal legal document to conclude the disputes. Often a consent document is used to signify the commitment of the parties to the terms of the settlement, but it is unclear what legal status such a document has. In particular, it is not certain that the document can bind the members of the group to follow the promises of the bargaining representative.

A formal process for resolving development disputes through negotiation necessitates the design of a document that has a certain legal status. Until there is one, bargainers will never be certain of what an agreement actually means.

How Can One Incorporate Technical Knowledge into This Review Process?

Technical knowledge can help a negotiation process to work efficiently and to produce decisions that insure environmental quality and public safety. At some point, designers will want to incorporate expert knowledge into the negotiation sessions. This knowledge may help the negotiators reach a settlement that is realistic, and it may widen the range of potential settlements considered. In addition, society will likely desire a finer technical scrutiny of the negotiated settlement to insure that existing environmental and safety norms will not be compromised in negotiation.

The trick for those designing a negotiation process is to develop a way to bring information into negotiations without creating conditions that undermine its acceptance. This task is complicated by the fact the developer is likely to have access to high levels of expertise concerning his or her own project, while others will have much less information. Since asymmetries of information can have a major negative impact on bargaining, proponents of negotiation will need to get information into discussions in a way that does not aggravate existing imbalances. In some disputes, such as the Snoqualmie Dam dispute, which involved regional environmental groups opposing local proponents of the dam, the developer can serve as an expert. In this particular dispute, the Army Corps of Engineers was able to offer advice to the disputants concerning different alternative projects. This conflict, however, was quite unusual in that the developer was not central to the overall dispute.

Generally, the developer will not be able to credibly provide information to the bargainers. Furthermore, environmental impact statements prepared by project proponents often fail to provide the needed information (O'Hare, Bacow, and Sanderson, 1983). More likely, the negotiation process will have to provide some financial support to those lacking it to gather the information that they want. When needed, this subsidy of the negotiation process may greatly assist the search for agreement. Instead of subsidizing contention (as is currently done), this siting process would subsidize information.

Conclusion

There is wide room for legislators or other designers of a negotiation-review process to tailor it to local conditions. Nevertheless, all design processes must address a similar set of issues. Since large projects often have substantial impacts on surrounding communities, review costs account for a small proportion of total costs, and uncertainty creates large penalties, these may prove especially appropriate for negotiated reviews. Similarly, projects sponsored by government or licensed monopolies, such as electric utilities, may seem particularly appropriate for review since public participation in these decisions has a long history.

Those designing negotiation processes must determine what issues should constitute the bargaining agenda. In making this determination, legislators should keep in mind that to a point, an increasing number of issues improves the opportunities for finding agreements that permit joint gains and cooperative searches for settlements. When there are few issues in the bargaining, negotiations can become competitive zero-sum battles. Without some formal designation of an appropriate bargaining agenda, the negotiators will remain subject to charges of shakedowns, sellouts, and power grabs.

Designers should determine a procedure for selecting bargaining participants. Participants should include the developer and public officials who have responsibility to review the project. A separate process, such as a petition, could determine who would represent the affected community and how they would be held accountable for their actions. A mediator should also participate. Because the disputing groups will likely meet for the first time during these negotiation sessions, the mediator could perform a valuable service by acting to resolve procedural questions and by chairing meetings. A series of deadlines can limit the potential for delay that negotiations might otherwise bring to a siting process.

Prudent designers must plan actions in the event of the failure of bargaining to settle a dispute. Negotiation theory suggests that this will be the largest determinant of negotiation power. The history of industrial relations indicates that achieving a balance is very much a political choice to which analysis has little to offer. Over time, as the results of a bargaining process become known, legislators can alter power balances by changing procedures for resolving impasses. Negotiations must conclude with a document that has legal stature and commits the signees to a course of action. All negotiated settlements, in addition, should face the technical scrutiny of a regulatory agency. This review, however, would focus less on the choice of site than on the appropriateness of the project plan and the impact of operations.

To date, there has been little experience with such a process. The Massachusetts Hazardous Waste Facility Siting Act has, however, incor-

porated many of the features suggested here. It restricts review to hazardous-waste facilities but approves a wide variety of mechanisms for
compensating a local community and suggests a fairly large issue agenda.
The law determines who participates in bargaining. The Massachusetts
act relies on that state's strong tradition of town government. Town officials both bargain on behalf of the community and can make appoint-
ments to complete a bargaining team (Mass. Gen. Laws, ch. 508, §15). The
act also enables the local community to get funds for preparing technical
reports to assist in bargaining. A siting council is given broad powers to
oversee the process. Agreements are given legal status and can be
enforced by the courts. Final plans, however, are subject to environmental and safety reviews by the state regulatory agency.

Perhaps the most interesting aspects of this act are the ways in
which it allocates power between the community, the developer, and a
special siting council. If 60 days after the preparation of a preliminary
project-impact report no agreement is reached, either the developer or
the local committee may appeal to the siting council for formal arbitration procedures (Mass. Gen. Laws, ch. 508, §5). This both creates a deadline to spur bargaining progress and enables parties to continue bargaining if neither side appeals. In addition, it determines what happens if
the disputants fail to reach a settlement. The siting council has 30 days
to make a formal determination of impasse and to set up an arbitration
procedure. (If both the community and the developer agree, this decision
may be delayed.) The arbitration panel will consist of an arbitrator chosen by the committee from the host community, an arbitrator chosen by
the developer, and an impartial arbitrator chosen by both. (If both the
community and the developer agree, a single arbitrator can be designated; §15.) If the parties cannot agree, the siting council can complete the arbitration panel. This process limits the power of the local
community to exclude it from consideration as a hazardous-waste
site but gives the community substantial power to determine the extent and types of compensation that it should get for hosting a waste
facility.

The recent passage of this law (1980) has made the experience
with it lean. O'Hare, Bacow, and Sanderson (1983) report that as of 1982,
three projects had initiated the process. One had advanced to the negotiation stage, but two had spawned legal battles concerning aspects of
the law. In particular, the act calls for the siting council to make a determination whether a site is feasible and deserving within 15 days of application by a developer. The act, however, is vague on the standard of review, and one town is suing on this issue. Another town has passed a

bylaw prohibiting a hazardous-waste facility within its boundaries. Only time will tell how the political and legal factors will work themselves out.

CHAPTER 11

Conclusion: Prospects for Negotiation

Introduction

As this book has discussed the advantages of resolving disputes over development projects through negotiations, it has raised a large number of cautions. Negotiations could become a series of hurdles that entrap developers in endless discussions. It is difficult to determine who should participate, what rights they have, and what constitutes an appropriate agenda. International experience warns that treaties and agreements do not insure that promises will be kept. Domestic labor–management experience indicates that even with the support of law and government, human interaction can turn bitter, adversarial, and destructive.

This chapter presents a comparison of three processes for addressing development disputes:

1. *Traditional siting process.* A developer acquires a site and completes plans for a potential project. Next, the developer announces the decision and begins applying for the needed federal, state, and local permits. Individuals and groups that desire to participate in the review of the project use public hearings and litigation to express their views. Effective participation requires the use of lawyers. Adversarial interaction between the developer, the regulatory agencies, and the opponents to a project commonly characterizes the process.

2. *Ad hoc negotiations.* Ad hoc negotiations supplement the traditional dispute process. At some point in the traditional process, a dispute reaches an impasse. The participants, usually with the

assistance of a mediator, then attempt to discuss the issues and reach a consensus. The selection of participants, negotiation agenda, deadlines, and bargaining procedures is set through prenegotiation conferences or resolved in an ad hoc way. When successful, the negotiators commit themselves to an agreement whose legal status varies with the dispute. When unsuccessful, disputants continue to contend within the traditional process of public hearings and judicial appeals.

3. *Formal negotiations.* A formal negotiation process would define a legitimate negotiation agenda, establish rules for selecting participants, use a mediator to resolve procedural issues, and set deadlines to encourage progress. Unlike in the other systems, there are really no existing examples of the use of this dispute-resolution process. A system of formal negotiations would set procedures to resolve disputes when no agreement was reached. They could include either arbitration of the developer's and the opposing group's positions or a vote in the community on the developer's last proposal. Some formal procedures would ratify settlements that would be legally binding. Technical experts would assist participants. Finally, a formal review of negotiated settlements would insure that existing standards and safeguards would be protected.

Process and Outcome Objectives

Any procedure for reviewing proposed development projects should meet a series of process and outcome objectives if it is to offer a viable alternative or complement to existing procedures. Although individuals would disagree on the importance of each objective, there already exists wide agreement concerning which objectives such a process should meet. Susskind (1981), Fisher (1979), and Cormick (1982) propose objectives and criteria for assessing different negotiation processes. In my view, the criteria that they have developed apply not just to negotiation processes, but to any method for balancing the interests of different groups. Sullivan (1980) and O'Hare et al. (1983) have developed objectives and criteria for the assessing site review processes. Combining these different perspectives, one can propose three different objectives for the site review process:

1. *Fairness.* The siting process should both *be fair* and *be perceived as fair*.

2. *Process efficiency.* The review process should itself be efficient.
3. *Producing outcomes that maximize social benefits.* The site selected and design chosen should allow the project to operate efficiently and not impose unreasonable risks on workers or the community.

Although these objectives compete with each other, they do set the framework of discourse for determining appropriate siting policies.

Fairness

Designing a siting process that uses fair procedures and produces a fair outcome is difficult. In particular, whether a process is perceived as fair is clearly beyond the control of its designers. The good intentions of the participants in siting reviews cannot insure that a community will consider the procedure fair. To succeed at all, the designers must scrutinize the procedures used to certify individuals as participants, to set the discussion agenda, and to resolve disputes over bargaining procedures.

1. Does the review process facilitate public participation?

The rules for participation and the rights given to the participants will greatly influence whether a community perceives a process as fair. Wide access to participation in a decision may help to insure it against charges of special-interest decision-making. The traditional siting process clearly defines who participates. Legislative efforts of the last decade have helped to incorporate public hearings into traditional proceedings, and judicial rulings have expanded the grounds for participating in review proceedings. For negotiations, unless some formal process for selecting negotiation participants and defining their rights is developed, then this process will raise doubts of its fairness. A negotiation process that restricts participation more narrowly than current practice is more likely to be characterized as unfair.

2. Is the decision or settlement consistent with preexisting practices?

Susskind (1981) and Fisher (1979) suggest that a negotiation settlement should be consistent with preexisting practices. Since preexisting practices will determine the range of a bargainer's expectations, the predictability of settlements will prove important to the perceived fairness of a siting process.

3. Are the results acceptable to the parties?

Insuring that the results of a negotiation process will be accepted as fair is difficult. Susskind (1981) and Fisher (1979) suggest that a negotiation settlement be acceptable to the parties in a dispute. This seems a particularly appropriate way for judging the fairness of any site review process.

4. Does the process improve relations between the parties?

Susskind (1981) and Fisher (1979) suggest that negotiation processes should help improve relations between the parties. This is likely an essential element in the public's perception of the fairness of a dispute resolution process. Both traditional and alternative siting processes should meet this criterion.

Efficiency of Review Process

Economic considerations argue that a siting process should operate in an efficient fashion. The project sited will often produce goods and services that must compete for a consumer's dollar.

5. Does the process produce quick, low-cost decisions?

Perhaps the critical factor in any review process is the time and effort of those who participate. In particular, siting processes should not require large sacrifices of the time of volunteers, or the labor of developers and government agencies.

Producing Outcomes That Maximize Social Benefits

The decision of any private or public process should strive to increase the net benefits to society. One way to do this is to insure that the development decision will reflect not only the costs that a developer bears, but also the costs that a decision imposes on others. This concern necessitates special attention to the safety of a project's design and site.

6. Does the process reconcile the interests of the parties in ways that leave no further opportunities for mutual gain?

When siting a project or negotiating a development design, participants should strive to make sure that no opportunities exist for further gains. In particular, they should seek to insure that there is no other way of making *all* participants better off through some alteration of project design or some alternative package of compensation.

7. Does the decision or settlement set a good precedent for the future?

The decision of a siting review process does not stand alone; it sets both expectations and standards for the future. Thus, it is important that a siting or negotiation process look not only at the issues that it faces at the immediate time, but also at those that its decisions pose.

8. Does the outcome protect public health and safety?

The choice of a site affects the overall safety of a facility's operations and its impact on the environment. Safety requires the selection of a site where natural or manmade events will not cause excessive accidents, and where the consequences of an accident, if one occurs, will be reduced.

Assessment

Those designing public policy do not get to choose the best of all possible worlds; rather, they choose in a world where uncertainty rules the day, good programs fail for unknown reasons, and poor programs succeed without any obvious cause. Choice requires comparison. In comparing formal negotiations with traditional adjudicatory proceedings and ad hoc forms of negotiation, one can see both the promise and the uncertainty that they hold.

One way to help insure the perceived fairness of a process is to create opportunities for individuals to participate. Administrative agencies and environmental law have expanded the abilities of individual citizens and groups to intervene in traditional siting reviews. This practice, however, has serious limitations. These traditional forms permit participation only in the very latest stages of the review process, where it often has little constructive effect. Issues raised late in the process can delay a decision and create major risks to a developer. Furthermore, forceful participation requires the development of a technical legal case. Although laws provide for recovery of legal fees, the traditional process bars nonlawyers from full participation. Ad hoc negotiations attempt to channel the energy and concern of individuals into a constructive form of interaction that does not require lawyers. Unfortunately, ad hoc bargaining uses the traditional processes to define the dispute, the issues, and the parties.

Bargaining starts only after a stalemate is reached. Participation comes late in the decision process, and the dispute is shaped by judicial tests. The price of participation is developing a legal position strong enough to halt the project in the traditional process. A formal negotiation process, however, should facilitate early participation in a decision process. No legal contest is necessary. Thus formal negotiation may avoid adversarial legal contests and facilitate constructive participation in the project's design.

Consistency with past practices and expectations is a key element in any concept of fairness. The heart of the traditional adjudicatory procedures is a concern for past precedent. In ad hoc negotiations, it is less sure that precedents will be respected. This is particularly true because the negotiation process is often seen as part of a unique set of activities. Participants generally meet for the first and only time in the beginning. In more formal negotiation processes, participants often look to other situations to see what guidelines or settlement patterns they can follow. Although concern for precedent is likely weaker in the traditional process, it may prove stronger than that in ad hoc bargaining.

For people to continue to perceive a process as fair, the results need to be acceptable to them. The traditional process has the weight of historical precedent behind it, as well as the force of the state. People have little choice other than to accept the process. Unfortunately the traditional process often requires developers to devote large sums of money to projects before getting a final decision. The reliance on expensive and legalistic forms of dispute resolution often forces groups to raise issues that have a powerful legal status rather than those that they care about. Both factors raise doubts over the fairness in practice of the traditional process.

An ad hoc negotiation process, by its very nature, produces outcomes that are acceptable to the participants. If no agreement is reached, the dispute reverts to the traditional process. Those who feel that either the process or the outcome of negotiations is unfair can appeal to the courts to scrutinize the decision. Ad hoc negotiations, however, remain vulnerable to the charges that the selection of participants, agenda of issues, and settlement terms produces cooperation that is inappropriate. For it to work, though, a good proportion of the disputes need to end with some form of a negotiated settlement that the parties can accept.

In formal negotiations, law would legitimate the choice of bargainers and agenda. Law would also constrain the settlements to ranges acceptable by a wider public. The fairness of the outcome will depend on how a formal negotiation process balances the interests of competing concerns.

For a long-term success, a dispute resolution process should improve the relations between the parties. One of the major successes of the traditional process is that it eventually does resolve a dispute. Procedures and laws are interposed between the disputants. Unfortunately, as the slow process of law grinds on, it can exacerbate the antagonisms between the parties. Ad hoc negotiations offer a potential process for allowing individuals to work with each other to resolve their differences in a way that often forces a debunking of dehumanizing stereotypes. Formal negotiations also share this good chance of leading to improved relations.

For a siting process to operate efficiently, it must produce quick, low cost, and stable decisions. The traditional process appears to fail to meet this standard. In particular, protracted litigation raises the costs of this decision process, and the uncertainty that it introduces into development planning makes it an unattractive alternative. Ad hoc negotiations can reduce the uncertainties that often plague that traditional process. However, the threat of litigation by a discontented individual remains.

A formal negotiation process is likely to require more resources initially. Participants will need access to technical information. The negotiation process will require time and preparation, as well as the heavy leadership and organizational costs outlined throughout this book. This process, however, may most effectively reduce the levels of delay and uncertainty that now plague the traditional process.

The site choice, project design, and mitigation measures should not possess defects that could be readily corrected. In particular, it should not be possible to make everyone better off and no one worse off through some modification of the design, site, or compensation package. It is in this dimension that the traditional process is particularly weak. By forcing individuals to dispute issues that are only surrogates for their true concerns, adjudication can result in projects meeting very strict emission standards when almost all would have preferred a settlement that addressed socioeconomic impacts.

Both ad hoc negotiations and formal negotiations enable the participants to discuss the issues that are of concern to them. The ability to reach an efficient settlement depends heavily on the skills of the bargainers and on the scope of the issue agenda. In ad hoc negotiations, these skills will likely prove lacking since organizations and leaders often mobilize for the adversarial context, rather than for discussion or bargaining. Formal negotiations may surmount this problem.

Finally, no process is really effective if it fails to provide a good precedent for the future and to adequately protect environmental and safety standards. The concern for precedent and safety marks the traditional process. Ad hoc negotiations generally take place within the traditional process. Although it is difficult to tell whether the final decision will produce a good precedent for future action, the safety and health standards generally remain exogenous to the negotiation. In a formal negotiation-review process, the negotiation itself will be seen as precedent-setting, which can help insure a special concern for precedent. As in ad hoc negotiations, external health and safety review can protect the concerns of the public.

Conclusion

Table 6 summarizes the comparison of traditional project-review processes, ad hoc negotiations, and formal negotiations. Each process has some advantages and disadvantages. The traditional process possesses the strengths and weaknesses of adjudication. It has a strong concern for precedent and uses institutions that possess political legitimacy. Unfortunately, the law limits the ability of participants to address the multiple issues raised by negotiations. The technical character of its decision-making makes direct participation by affected communities difficult.

Ad hoc negotiations overcome some of these rigidities but raise questions concerning the legitimacy of the bargaining process. In particular, difficult questions arise over the designation of participants, the legitimacy of the issues on the agenda, and the legal status of the settlements. This process, however, has proved effective when a dispute's issues are well defined and the disputants desire to surmount stalemates.

Although useful in many situations, ad hoc negotiations are likely to serve as a complement to adjudication only in those rare situations where groups, issues, and leaders all support a negotiated settlement. Without the support of law, the ad hoc efforts to resolve development disputes will follow the model of international relations. The bargainers will continue to have unclear rights, and the settlements will continue to possess an uncertain legal status. As in international relations, only weak incentives will discourage destructive forms of conflict. Perhaps most importantly, the mobilization and organizational efforts needed to win recognition and bargaining power in development disputes will likely produce leaders and groups that lack the skills and inclination to bargain. This common failing will prove especially important, for, as in international relations, bargaining can succeed only when leaders will bear the costs and criticism that arise in articulating a group's interests and considering alternative positions and perspectives. As long as laws continue to finance successful litigation, yet provide no support for lead-

Table 6
Assessment of Site Review Processes

Criteria	Traditional negotiations	Ad hoc negotiations	Formal negotiations
Fairness			
1. Does the review process facilitate public participation?	Yes, but participation requires the use of lawyers.	Yes, but it is never certain that all interests are represented.	Yes. Law can also legitimate the participation of groups.
2. Is the decision or settlement consistent with preexisting practices?	Strength of traditional process is its concern for precedent.	The high risk of inconsistent settlements is moderated by judicial scrutiny.	Formal procedures and limited judicial scrutiny control the risk of decisions inconsistent with past practice.
3. Are the results acceptable to the public?	Once legal challenges are exhausted, disputants *must* accept decision.	No settlement exists unless a consensus is reached. Often, no settlement will be reached.	Negotiated settlements will share a consensus. Those resolved otherwise will carry legal force.
4. Does the process improve relations between the parties?	Adversarial character often exacerbates antagonisms.	Generally improves relations.	Good chance of improving relations.
Efficiency of review process			
5. Does the process produce quick, low-cost decisions?	Process has proved costly and uncertain.	Has worked well in stalemated disputes, but ad hoc character may limit wider use. Can become a tactic for delay.	Uncertain, but the incorporation of formal structures, such as deadlines, may help performance.
6. Does the process reconcile the interests of the parties in ways that leave no opportunities for mutual gain?	Legal rigidities often preclude the maximization of joint gains.	Achievement of efficient outcomes depends on skills of participants. These may prove a key limiting factor.	Achievement of efficient outcomes depends on skills of participants and scope of the bargaining agenda.
7. Does the decision or settlement set a good precedent for the future?	Very likely.	Hard to predict. Very likely risky.	Less likely to set good precedent than traditional process, but more likely than ad hoc bargaining.
8. Does the outcome protect public health and safety?	Existing standards are enforced.	Existing standards are enforced.	Depends on the incorporation of expertise into bargaining and on the adequacy of postnegotiation reviews.

ers to follow the more costly and difficult path of negotiation, one must have a pessimistic assessment of ad hoc negotiations.

The potential of a *formal* process for resolving development disputes through negotiation is much greater. Just as labor law works to set bargaining units and to regulate industrial conflict, the law can help efforts to resolve development disputes by determining who should bargain and what rights and duties they have. The law can balance the powers of each party in a dispute. The law can support negotiated agreements. In particular, the law can limit the ability of any one group or individual to delay development projects, while offering an opportunity to influence the design of a project and the types of compensation offered. Most importantly, the law and public policy can legitimate the effort to find a more constructive means of resolving disputes.

References

Ackerman, Bruce, and William T. Hassler. *Clean Coal/Dirty Air.* New Haven, Conn.: Yale University Press, 1981.

Alinsky, Saul D. *Rules for Radicals: A Practical Primer for Realistic Radicals.* New York: Random House, 1971.

Bacow, Lawrence. "Exploring Environmental Impacts: Beyond Quantity and Quality." *Technology Review,* Vol. 85, No. 1 (Jan. 1982), 32–37.

Bacow, Lawrence, and James R. Milkey. "Overcoming Local Opposition to Hazardous Waste Facilities: The Massachusetts Approach." *Harvard Environmental Law Review,* Vol. 6 (Spring 1982), 265–305.

Bardach, Eugene, and Robert Kagan. *Going by the Book.* Philadelphia: Temple University Press, 1982.

Bardach, Eugene, and Lucian Pugliaresi. "The Environmental Impact Statement versus the Real World." *The Public Interest* (Fall 1977), 22–38.

Barton, John H. "Behind the Legal Explosion." *Stanford Law Review* Vol. 27, (Feb. 1975), 567–584.

Bartunek, Jean M., Alan A. Benton, and Christopher B. Keys. "Third Party Intervention and the Bargaining Behavior of Group Representatives." *Journal of Conflict Resolution,* Vol. 19 (1975), 523–557.

Bernstein, Marver H. *Regulating Business by Independent Commission.* Princeton, N.J.: Princeton University Press, 1955.

Bok, Derek, and John Dunlop. *Labor and the American Community.* New York: Simon & Schuster, 1970.

Burgess, Heidi, and Douglas Smith. "The Uses of Mediation." In Susskind, Lawrence, Lawrence Bacow, and Michael Wheeler, eds., *Resolving Environmental Regulatory Disputes.* Cambridge, Mass.: Schenkman, 1984.

Burton, John W. *Conflict and Communication.* New York: Macmillan, 1969.

Carter, Jimmy. *Keeping Faith: Memoirs of a President.* New York: Bantam Books, 1982.

Cohen, Herb. *You Can Negotiate Anything.* New York: Bantam Books, 1980.

Congressional Record. Vol. 23, No. 170 (Oct. 20, 1977), 34547–34559.

Cormick, Gerald W. "The 'Theory' and Practice of Environmental Mediation." *The Environmental Professional,* Vol. 2 (1980), 24–33.

Cormick, Gerald W. "Intervention and Self-Determination in Environmental Disputes: A Mediator's Perspective." *Resolve* (Winter 1982), 1–3.

Cormick, Gerald W., and Jane McCarthy. *Environmental Mediation: First Dispute.* Seattle: Office of Environmental Mediation, University of Washington, 1974.

Cormick, Gerald W., and Leota Patton. *Environmental Mediation: Defining the Process through Experience.* Seattle: Office of Environmental Mediation, University of Washington, 1977.

Deutsch, Morton. *The Resolution of Conflict.* New Haven, Conn.: Yale University Press, 1973.

Douglas, Ann. "The Peaceful Settlement of Industrial and Intergroup Disputes." *Journal of Conflict Resolution,* Vol. 1 (1957), 69–81.

Douglas, Ann. *Industrial Peacemaking.* New York: Columbia University Press, 1962.

Douglas, Mary, and Aaron Wildavsky. *Risk and Culture.* Berkeley: University of California Press, 1982.

Druckman, Daniel, and Robert Mahoney. "Processes and Consequences of International Negotiations." *Journal of Social Issues,* Vol. 33 (1977), 60–87.

Ducsik, Dennis W. *Electricity Planning and the Environment: Towards a New Role for Government in the Decision Process.* Unpublished Ph.D. dissertation. M.I.T., 1978.

Dunlop, John. *Industrial Relations Systems.* New York: Holt, Rinehart & Winston, 1958.

Environmental Mediation: An Effective Alternative? A report of a conference held in Reston, Virginia, Jan. 11–13, 1978. Palo Alto, Calif.: Resolve, Center for Environmental Conflict Resolution, 1978.

Finer, Samuel E. *The Man on Horseback: The Role of the Military in Politics.* Harmondsworth, England: Penguin Books, 1975.

Fisher, Roger. *International Conflict for Beginners.* New York: Harper and Row, 1969.

Fisher, Roger. *Some Notes on Criteria for Judging the Negotiation Process.* Unpublished paper distributed at the Negotiations Seminar of the Harvard Negotiation Project, Harvard Law School, Nov. 1979.

Fisher, Roger, and William Ury. *Getting to Yes.* Boston: Houghton-Mifflin Company, 1981.

Freedman, James O. *Crisis and Legitimacy: The Administrative Process and American Government.* New York: Cambridge University Press, 1978.

Frieden, Bernard J. *The Environmental Protection Hustle.* Cambridge: M.I.T. Press, 1979.

Friedman, Frank B. "Environmental Checklist." *Real Property Probate and Trust Journal,* Vol. 14 (1979), 873–884.

Fuller, Lon L. "Mediation—Its Forms and Functions." *Southern California Law Review,* Vol. 44 (1971), 305–339.

Fuller, Lon L. "The Forms and Limits of Adjudication." *Harvard Law Review,* Vol. 92 (1978), 353–409.

Hamilton, Alexander, James Madison, and John Jay. *The Federalist Papers.* New York: New American Library, 1961.

Harsanyi, John C. "Bargaining in Ignorance of the Opponent's Utility Function." *Journal of Conflict Resolution,* Vol. 6, No.1 (1972), 29–38.

Hastings, Max, and Simon Jenkins. *The Battle for the Falklands.* New York: Norton, 1983.

Herman, E. Edward, and Alfred Kuhn. *Collective Bargaining and Labor Relations.* Englewood Cliffs, N.J.: Prentice-Hall, 1981.

Hirschman, Albert O. *Exit, Voice and Loyalty.* Cambridge: Harvard University Press, 1970.

Homer. *The Iliad,* translated by Robert Fitzgerald. New York: Anchor Books, 1975.

Iklé, Fred C. *How Nations Negotiate.* New York: Harper and Row, 1964.

Ilich, John. *Power Negotiating.* New York: Playboy Paperbacks, 1980.

Jackson, Elmore. *Meeting of Minds: A Way to Peace through Mediation.* New York: McGraw-Hill, 1952.

Jervis, Robert. *The Logic of Images in International Relations.* Princeton, N.J.: Princeton University Press, 1970.

Karrass, Chester L. *Give and Take: The Complete Guide to Negotiating Strategies and Tactics.* New York: Thomas Y. Crowell, 1974.

Keesing's Contemporary Archives, Vol. 28 (1982), 31709–31718.

Kelley Blue Book. Costa Mesa, Calif.: Kelley Blue Book Company, 1984.

Kelley, Harold H. "A classroom study of dilemmas in interpersonal negotiations." In Kathleen Archibald, ed., *Strategic Interaction and Conflict.* Berkeley: Institute of International Studies, University of California, 1966.

Kerr, Clark. "Industrial Conflict and its Mediation." *American Journal of Sociology,* Vol. 60, No. 3 (1954), 230–245.

Kissinger, Henry. *American Foreign Policy,* 3rd. ed. New York: Norton, 1977.

Kissinger, Henry. *White House Years.* Boston: Little, Brown, 1979.

Kissinger, Henry. *Years of Upheaval.* Boston: Little, Brown, 1982.

Kolko, Gabriel. *The Triumph of Conservatism.* New York: The Free Press of Glencoe, 1963.

Kretzmer, David. *Legal Problems of Binding Communities to Compensation Agreements for Adverse Effects of Energy Facilities.* Cambridge: Laboratory of Architecture and Planning, Massachusetts Institute of Technology, 1979.

Landsberger, Henry A. "Interim Report on a Research Project in Mediation." *Labor Law Journal,* Vol. 6 (Aug. 1955), 552–560.

Landsberger, Henry A. "A Final Report on a Research Project in Mediation." *Labor Law Journal,* Vol. 7 (Aug. 1956), 501–510.

Livernash, E. Robert. *Collective Bargaining in the Basic Steel Industry.* Washington: U.S. Government Printing Office, 1961. For the U.S. Department of Labor.

Massachusetts General Laws, ch. 508.

Mazmanian, Daniel A., and Jeanne Nienaber. *Can Organizations Change?* Washington: Brookings Institution, 1979.

McKersie, Robert, Charles R. Perry, and Richard Walton. "Intraorganizational Bargaining in Labor Negotiations." *Journal of Conflict Resolution,* Vol. 9, No. 4 (Dec. 1965), 463–481.

McNaughton, Wayne, and Joseph Lazar. *Industrial Relations and the Government.* New York: McGraw-Hill, 1954.

Memorandum of Agreement between the Governments of the United States of America and the State of Israel. March 26, 1979.

Memorandum of Agreement between the Governments of the United States and Israel—Oil. March 26, 1979.

Mernitz, Scott. *Mediation of Environmental Disputes: A Sourcebook.* New York: Praeger, 1980.

National Labor Relations Act. 29 USC §158.

Neustadt, Richard E. *Alliance Politics.* New York: Columbia University Press, 1970.

"New England Power Says Plans Canceled for Nuclear Facility." *Wall Street Journal* (Western Edition), Dec. 18, 1979, p.11.

Nozick, Robert. *Anarchy, State, and Utopia.* New York: Basic Books, 1974.

O'Hare, Michael. "'Not on My Block You Don't': Facility Siting and the Strategic Importance of Compensation." *Public Policy,* Vol. 25 (1977), 407–458.

O'Hare, Michael, Lawrence Bacow, and Debra Sanderson. *Facility Siting and Public Opposition.* New York: Van Nostrand Reinhold, 1983.

Olson, Mancur. *The Logic of Collective Action.* Cambridge: Harvard University Press, 1973.

President's Committee on Administrative Management. *Report of the Committee with Studies of Administration in Federal Government*. Washington: U.S. Government Printing Office, 1937.

Pruitt, Dean G. "Indirect Communication and the Search for Agreement in Negotiation." *Journal of Applied Social Psychology*, Vol. 1, (1971), 204–239.

Pruitt, Dean G. *Negotiation Behavior*. New York: Academic Press, 1981.

Pruitt, Dean G., and Douglas F. Johnson. "Mediation as an Aid to Face-Saving in Negotiation." *Journal of Personality and Social Psychology*, Vol. 14 (1970), 239–246.

Raiffa, Howard. *Decision Analysis: Introductory Lectures on Choices under Uncertainty*. Reading, Mass.: Addison-Wesley, 1968.

Raiffa, Howard. *The Art and Science of Negotiation*. Cambridge: Harvard University Press, 1982.

Rawls, John. *A Theory of Justice*. Cambridge: Harvard University Press, 1971.

Rifkind, Simon H. "Are We Asking Too Much of Our Courts?" *70 F.R.D. 79, 96*. The Pound Conference, 1976.

Rivkin, Malcolm. *An Issue Report: Negotiated Development, A Breakthrough in Environmental Controversies*. The Conservation Foundation, 1977.

Rubin, Jeffrey, and Bert Brown. *The Social Psychology of Bargaining and Negotiation*. New York: Academic Press, 1975.

Salinger, Pierre. *America Held Hostage: The Secret Negotiations*. Garden City, N.Y.: Doubleday, 1981.

Sander, Frank E. "Varieties of Dispute Resolution." *70 F.R.D. 79*. The Pound Conference, 1976.

Schelling, Thomas C. *The Strategy of Conflict*. New York: Oxford University Press, 1960.

Shapiro, Fred C. "Mediator." *New Yorker*, Vol. 46 (1970), 36–58.

Simkin, William E. "Code of Professional Conduct for Labor Mediators." *Labor Law Journal*, Vol. 15, No. 10 (Oct. 1964), 627–633.

Simkin, William. *Mediation and the Dynamics of Collective Bargaining*. Washington: The Bureau of National Affairs, 1971.

Slichter, Sumner H., James J. Healy, and E. Robert Livernash. *The Impact of Collective Bargaining on Management*. Washington: Brookings Institution, 1975.

Sobel, Lester A., ed. *Peace-Making in the Middle East*. New York: Facts on File, 1980, 235.

Stenelo, Lars G. *Mediation in International Negotiations*. Sweden: Studentilleratur, 1972 (Lund Political Studies #14).

Stevens, Carl M. *Strategy and Collective Bargaining Negotiations*. New York: McGraw-Hill, 1963.

Stewart, Richard. "The Reformation of American Administrative Law." *Harvard Law Review*, Vol. 88, No. 8 (June 1975), 1667–1813.

Stigler, George. "The Theory of Economic Regulation." *Bell Journal of Economic and Management Science*, Vol. 2 (1971), 3–21.

Sullivan, Timothy J. *Negotiation-Based Review Processes for Facility Siting*. Unpublished Ph.D. dissertation. Harvard University, 1980.

Sullivan, Timothy J. "The Difficulties of Mandatory Negotiation (The Colstrip Power Plant Case)." In Susskind, Lawrence, Lawrence Bacow, and Michael Wheeler, eds., *Resolving Environmental Regulatory Disputes*. Cambridge, Mass.: Schenkman, 1984.

Summers, Clyde W. "Ratification of Agreements." In Dunlop, John T., and Neil W. Chamberlain, eds., *Frontiers of Collective Bargaining*. New York: Harper and Row, 1967.

Susskind, Lawrence. "Environmental Mediation and the Accountability Problem." *Vermont Law Review*, Vol. 6, No. 1 (Spring 1981), 1–48.

Susskind, Lawrence, and Alan Weinstein. "Towards a Theory of Environmental Dispute Resolution." *Boston College Environmental Affairs Law Review*, Vol. 9, No. 2 (1980), 311–357.

Susskind, Lawrence E., James R. Richardson, and Kathryn J. Hildebrand. *Resolving Environmental Disputes: Approaches to Intervention, Negotiation, and Conflict Resolution.* Cambridge: M.I.T., Environmental Impact Assessment Project, 1978.

Talbot, Allan. *Settling Things: Six Case Studies in Environmental Mediation.* Washington: The Conservation Foundation, 1983.

Treaty of Peace between the Arab Republic of Egypt and the State of Israel.

Tribe, Laurence. "Ways Not to Think about Plastic Trees." In Tribe, Laurence, Corinne S. Schelling, and John Voss, eds., *When Values Conflict: Essays on Environmental Analysis, Discourse, and Decision.* Cambridge, Mass.: Ballinger, 1976.

Truman, David B. *The Governmental Process: Political Interests and Public Opinion.* New York: Knopf, 1951.

Ullman, Lloyd. *The Rise of the National Trade Union.* Cambridge: Harvard University Press, 1955.

Wall, James A. "Mediation: An Analysis, Review, and Proposed Research." *Journal of Conflict Resolution*, Vol. 25, No. 1 (March 1981), 157–183.

Walton, Richard E. *Interpersonal Peacemaking: Confrontations and Third Party Consultation.* Reading, Mass.: Addison-Wesley, 1969.

Wilson, James Q. *Political Organizations.* New York: Basic Books, 1973.

Young, Oran R. *The Intermediaries: Third Parties in International Crises.* Princeton, N.J.: Princeton University Press, 1967.

Young, Oran R. *Bargaining.* Princeton, N.J.: Princeton Center of International Studies, Princeton University, 1970.

Zartman, I. William "Negotiation as a Joint Decisionmaking Process." *Journal of Conflict Resolution*, Vol. 21 (1977), 619–638.

Walton, Richard E. *Interpersonal Peacemaking: Confrontations and Third Party Consultation.* Reading, Mass.: Addison-Wesley, 1969.

Index